SpringerBriefs in Food, Health, and Nutrition

Springer Briefs in Food, Health, and Nutrition present concise summaries of cutting edge research and practical applications across a wide range of topics related to the field of food science, including its impact and relationship to health and nutrition. Subjects include:

- Food chemistry, including analytical methods; ingredient functionality; physic-chemical aspects; thermodynamics
- Food microbiology, including food safety; fermentation; foodborne pathogens; detection methods
- Food process engineering, including unit operations; mass transfer; heating, chilling and freezing; thermal and non-thermal processing, new technologies
- Food physics, including material science; rheology, chewing/mastication
- Food policy
- And applications to:
  - Sensory science
  - Packaging
  - Food quality
  - Product development

We are especially interested in how these areas impact or are related to health and nutrition.

Featuring compact volumes of 50 to 125 pages, the series covers a range of content from professional to academic. Typical topics might include:

- A timely report of state-of-the art analytical techniques
- A bridge between new research results, as published in journal articles, and a contextual literature review
- A snapshot of a hot or emerging topic
- An in-depth case study
- A presentation of core concepts that students must understand in order to make independent contributions

More information about this series at http://www.springer.com/series/10203

Meera Verma

Energy Use in Global Food Production

Considerations for Sustainable Food Security in the 21st Century

 Springer

Meera Verma
Headland Vision
  Adelaide
South Australia
Australia

ISSN 2197-571X   ISSN 2197-5728   (electronic)
SpringerBriefs in Food, Health, and Nutrition
ISBN 978-3-319-16780-0   ISBN 978-3-319-16781-7   (eBook)
DOI 10.1007/978-3-319-16781-7

Library of Congress Control Number: 2015934048

Springer Cham Heidelberg New York Dordrecht London

Springer International Publishing AG Switzerland is part of Springer Science+Business Media
(www.springer.com)

This work is dedicated to my mother Daphne Verma, and my mother-in-law Mary Headland, both of whom instilled in me the value of the limited resources on our Planet. Though neither of them used this terminology, instead they lived frugally by 'waste not, want not' and 'living within your budget', and still managed to have a lot of fun in life and exude a lot of love.

Acknowledgments

I would like to thank Michael Headland for his support while I compiled the data and wrote this book on top of a full consulting load, and my two anonymous reviewers for the valuable feedback they provided.

Contents

Abstract

This book examines the sustainability of energy use in global food production and consumption. Modern industrial agriculture uses fossil fuel, both directly to grow crops and indirectly to produce fertilizers, pesticides and farm machinery. Additional energy is used to transport and process food at a primary and secondary level. Thus food production is already a significant contributor to anthropogenic climate change. In addition, the median forecast for global population is more than 9.6 billion by 2050, a 33 % increase over the current population. Climate change predictions are that major food producing areas are likely to be impacted by extreme weather events and a warming world, with more frequent and deeper droughts and the threat of more invasive agricultural pests. Bioenergy and the use of biomass for production of fuel also have the potential to impact food production and arable land use. Together these forecasts have important considerations for global food production and food security. The nexus between food, water and energy are explored, against a background of climate change. Current efforts to reduce food loss and wastage, as well as improve the energy intensity of food and increase sustainability are also explored.

Keywords Food production · Food waste · Industrial agriculture · Climate change · Anthropogenic emissions · Energy intensity · Sustainable cropping · Conservation agriculture · Population growth

About the Author

Meera Verma Ph.D., FTSE, FAICD, is a professional executive with expertise spanning the global healthcare, biotechnology and renewable energy sectors, based in Adelaide, Australia. She is a professional non-executive director and the Principal of Headland Vision, providing strategic product development advisory services focused on biofuel development. She previously served as Site Director for the Adelaide-based R&D and manufacturing facility of Hospira Inc, a Chicago-based global specialty pharmaceutical and medication company.

Dr. Verma is a Fellow of both the Australian Academy of Technological Sciences and Engineering (ATSE) and the Australian Institute of Company Directors (AICD) and holds a Doctoral Degree in Biochemistry from the University of Adelaide. She is a non-Executive Director of Ellex Medical Ltd., Biosensis Pty Ltd and Trees for Life and is currently Chair of the SA Division of ATSE.

Dr. Verma trained as a Climate Reality Leader in 2014 and is associated with the group South Australian's for Climate Action.

Chapter 1
Introduction

The energy needs of a society are intrinsically tied to the production of food, and the efficiency of food production, among other things. Inability to balance these key factors will jeopardize food security and survival of that society.

Most people think of energy and food from the perspective of the embedded kilo joules (kJ) or kilo calories (kcal) available on consumption of that food. However, there is a much deeper connection and one that has been obscured over the last century by the availability of cheap fuel energy for use in the production of food.

Early civilizations relied on manual human and animal labour for the production of food, and wood or charcoal for cooking and preparing that food for human consumption. However, in medieval Europe, Britain began the use of fossil fuel in the form of coal, of which it had a plentiful supply as the country was sited on the "carboniferous crescent" from Scotland to the Ruhr (Christian 2009). Coal provided most of the energy required for British society, including that needed for producing and consuming food.

In the modern era, we use fossil fuel for multiple steps in the production of food. Energy from fossil fuel is used directly to cultivate crops or produce livestock and for processing the primary production into secondary or final product for consumption. Energy is also used indirectly for a number of inputs and processes. For example, in the production of fertilizers and pesticides; for accessing and supplying water for irrigation; for transportation of the food stock from the primary producer to the processor and then to wholesalers and retailers; and finally by the consumer to source, transport and prepare the food for consumption.

In the developed world, an average US farmer is estimated to use 3 kcal of fossil energy to produce 1 kcal of food energy. This increases to 35:1 for feedlot beef, not including the energy required for processing and transportation of the food (Horrigan et al. 2002). Horrigan and co-workers estimate that the food production system accounts for 17 % of all fossil fuel use in the United States.

© The Author(s) 2015
M. Verma, *Energy Use in Global Food Production*,
SpringerBriefs in Food, Health, and Nutrition, DOI 10.1007/978-3-319-16781-7_1

There is an increasing demand for food, driven by the projected growth in human population. During the years 2005–2010, the average global population growth rate was 1.20 % per annum. This rate is projected to decline to 0.51 % per annum by 2050 and to 0.11 % by 2100. However, despite the declining growth rate from the highs of >2 % in the 1960s, and assuming the decrease in fertility continues as projected, the median forecast for global population by 2050 is 9.6 billion and 10.9 billion by 2100 (United Nations 2013). An increase by the end of the century of 51 % over the 7.2 billion global population in mid-2013.

49 % of land surface area capable of supporting biomass and 70 % of extracted freshwater is already utilized for global food production. Demand for food is forecast to increase significantly with estimates that net global food production needs to be increased by 60–70 % by 2050. How is this demand to be met in a sustainable manner?

The Challenge

There has been an intensification of energy use in food production, and an increase in food demand, due largely to population growth and increasing affluence. Food production contributes to climate change, through both, use of fossil fuel and land use change (Woods et al. 2010). These factors have major implications for sustainable food security in the medium and longer term in the 21st century.

The Food and Agriculture Organization of the United Nations articulates the challenge succinctly in their report entitled "Energy-Smart Food at FAO" (FAO 2012) *"Our ability to reach food productivity targets may be limited in the future by a lack of inexpensive fossil fuels. This has serious implications both for countries that benefited from the initial green revolution and for those countries that are looking to modernize their agrifood systems along similar lines. Modernizing food and agriculture systems by increasing the use of fossil fuels as was done in the past may no longer be an affordable option. We need to rethink the role of energy when considering our options for improving food systems."*

By early 2015 the price of fossil oil dropped to its lowest level in 6 years due to a combination of overproduction, declining demand and slowing world-wide GDP growth (Krishnan 2015). This trend is forecast to continue with lower prices predicted to last through most of 2015 (Therramus and Austin 2015), (van der Hoeven 2014). However, the lower prices do not take into consideration externalities like greenhouse gas emissions (GHG) and their impact on climate change and the environment (IPCC 2014). As the world moves towards decarbonization, energy prices will change and this will undoubtedly have an impact on the cost of production and price of food.

The following chapters explore the food and energy connection, as well as the energy-food-water nexus in more detail. The impact of changing global demographics and climate change on food production and demand are also considered. Current loss of food at producer level and waste of food at consumer level is examined in the context of the lowest overall cost of meeting the expected demand for food. And finally examples of the emergence of sustainable intensification of agriculture, with attendant lower energy use are presented to offer a perspective on attaining sustainable, global food security.

References

Christian, D. 2009. Contingency, pattern and the S-curve in human history. *World history connected* retrieved 23 Jan 2015, from http://worldhistoryconnected.press.illinois.edu/6.3/christian.html.

FAO. 2012. Energy-smart food at FAO, food and agriculture organisation of the United Nations.

Horrigan, L., R.S. Lawrence, and P. Walker. 2002. How sustainable agriculture can address the environment and human health harms of industrial agriculture. *Environmental Health Perspectives* 110(5): 445.

IPCC. 2014. Climate change 2014 synthesis report. Intergovernmental Panel on Climate Change.

Krishnan, B. 2015. Oil falls again as IMF cuts forecast; Iran hints at \$25 oil. Retrieved 23 Jan 2015, from http://uk.reuters.com/article/2015/01/20/uk-markets-oil-idUKKBN0KT02W20150120.

Therramus, T., and S. Austin. 2015. Will collapse in oil price cause a stock market crash? From http://oil-price.net/en/articles/will-collapse-in-oil-price-cause-stock-market-crash.php.

United Nations, D. (2013). World population prospects: The 2012 revision, volume I: comprehensive tables ST/ESA/SER.A/336. New York: Department of Economic and Social Affairs, Population Division.

van der Hoeven, M. 2014. Cheap Oil's make-or-break moment for clean energy. http://www.huffingtonpost.com/maria-van-der-hoeven/cheap-oils-make-or-break_b_6307970.html 2015.

Woods, J., A. Williams, J.K. Hughes, M. Black, and R. Murphy. 2010. Energy and the food system. *Philosophical Transactions of the Royal Society B: Biological Sciences* 365(1554): 2991–3006.

Chapter 2
The Food and Energy Connection

As mentioned in the previous chapter, energy from fossil fuel is now used both directly and indirectly at multiple stages of the food production and supply chain, from primary production, processing, retailing, transportation and ultimately by the consumer. This relationship is illustrated by the following extract from the work of Canning and coworkers, using a hypothetical purchase of a non-organic salad mix by a consumer living on the East Coast of the US

> In this case, fresh vegetable farms in California harvest the produce to be used in the salad mix a few weeks prior to its purchase. The farms' fields are seeded months earlier with a precision seed planter operating as an attachment to a gasoline-powered farm tractor. Between planting and harvest, a diesel-powered broadcast spreader applies nitrogen-based fertilizers, pesticides, and herbicides, all manufactured using differing amounts of natural gas and electricity and shipped in diesel-powered trucks to a nearby farm supply wholesaler. Local farmers travel to the wholesaler in gasoline-powered vehicles to purchase farm supplies. The farms use electric-powered irrigation equipment throughout much of the growing period. At harvest, field workers pack harvested vegetables in boxes produced at a paper mill and load them in gasoline-powered trucks for shipment to a regional processing plant, where specialized machinery cleans, cuts, mixes, and packages the salad mixes. Utility services at the paper mill, plastic packaging manufacturers, and salad mix plants use energy to produce the boxes used at harvest and the packaging used at the processing plant, and for processing and packaging the fresh produce. The packaged salad mix is shipped in refrigerated containers by a combination of rail and truck to an East Coast grocery store, where it is placed in market displays under constant refrigeration.
>
> To purchase this packaged salad mix, a consumer likely travels by car or public transportation to a nearby grocery store. For those traveling by car, a portion of the consumer's automobile operational costs, and his or her associated energy-use requirements, help facilitate this food-related travel. At home, the consumer refrigerates the salad mix for a time before eating it. Subsequently, dishes and utensils used to eat the salad may be placed in a dishwasher for cleaning and reuse—adding to the electricity use of the consumer's household. Leftover salad may be partly grinded in a garbage disposal and washed away to a wastewater treatment facility, or disposed, collected, and hauled to a landfill (Canning et al. 2010).

© The Author(s) 2015
M. Verma, *Energy Use in Global Food Production*,
SpringerBriefs in Food, Health, and Nutrition, DOI 10.1007/978-3-319-16781-7_2

Energy Use in Primary Food Production

Direct fossil energy inputs into agriculture have generally been outweighed by yield improvements that deliver positive energy ratios, i.e., the energy content of the crop is greater than the energy utilized to produce that crop. However, when the embodied energy, i.e., energy utilized over the life cycle of the crop is considered, in some instances more energy can be used than is contained in the final product, as reviewed by Woods et al. (2010) for a range of crops in the UK. These studies used a standard 'cradle to grave' approach for life cycle assessment (LCA) of environmental impacts of a process or product. The review covered three field crops (bread wheat, oilseed rape and potatoes), four meats (beef, poultry, pork and lamb), milk and eggs, and tomatoes as the main protected crop. Apples and strawberries were also analyzed. Primary production to the farm gate was studied to provide the LCA.

For arable crops, energy inputs to produce the UK's main crops range from 1 to 6 GJ t^{-1} (Table 2.1). The authors examined conventional and organic farming methods, though direct comparisons methods can be problematic, since it appears that reduced direct use of fertilizers in the latter methods, is balanced out by lower yields. In general, oilseed rape is the highest energy consumer, given low yields and high fertilizer use, but the grain is more energy dense than cereals or legumes. For production of bread wheat, used as a proxy for cereals in the study, half the energy used is for fertilization, of which 90 % of the energy is in nitrogen production. Pesticide manufacture, by contrast, accounts for less than 10 % of the energy use. Potato cropping is more energy intensive than cereals and legumes due to the energy used for cool storage for long periods. However, because potatoes are a high yielding crop, they have a lower energy use per tonne harvested. Interestingly, if the energy use per tonne is calculated on a dry biomass basis, the energy intensity of potatoes is much higher since they contain 80 % water, compared to 15–20 % for wheat grain.

Although fossil fuels remain the dominant source of energy for agriculture, the mix of fuels used differs owing to the different fertilization and cultivation

Table 2.1 Primary energy used in arable crop production

	Primary energy used GJ t^{-1}	
	Non-organic	Organic
Bread wheat (UK)	2.52	2.15
Oilseed rape (UK)	5.32	6.00
Potatoes main crop (UK)	1.46	1.48
Feed wheat (UK)	2.32	2.08
Winter barley (UK)	2.43	2.33
Field beans (UK)	2.51	2.44
Soya beans (US)	3.67	3.23
Sugarcane (Brazil)	0.21	
Maize (US)	2.41	

Adapted from Woods et al. (2010)

Table 2.2 Energy used in animal production at commodity level in the UK

Commodity	Poultry	Pig meat	Beef	Lamb meat	Milk	Eggs
Unit	1 t ecw	1 t ecw	1 t ecw	1 t ecw	m^3	1 t
Primary energy (GJ)	17	23	30	22	2.7	12
Feed (%)	71	69	88	88	71	89
Manure and litter (%)	2	1	1	1	0	−4
Housing (%)	1	4	0	0	3	3
Direct energy (%)	25	26	11	11	26	12

ecw edible carcass weight (killing out percentage × live weight), slaughter not included
1 m^3 milk weighs about 1 tonne and 15,900 eggs weigh 1 tonne
Adapted from Woods et al. (2010)

requirements of individual crops and different approaches in different parts of the world (Woods et al. 2010). For example, in Europe, nitrogen fertilizer production uses large amounts of natural gas, and can account for more than 50 % of total energy use in commercial agriculture. Fossil oil accounts for between 30 and 75 % of energy inputs of UK agriculture, depending on the cropping system. However, in China 80 % of the energy for nitrogen fertilizer production comes from coal. In addition, in most regions, the embodied energy in farm machinery is an overhead of 40 % of diesel used for production.

The energy used per tonne of animal production is higher than that used for cropping (Woods et al. 2010), since animals are fed on crops and convert crop energy into higher quality protein and nutrients (Table 2.2). Feed provision is the dominant term in energy use (average of about 75 %). Direct energy use includes managing field stock, heating for young birds and piglets, and ventilation for pigs and poultry. Housing contributes a small fraction of total energy inputs, and is lower for more extensive systems, like free-range hens. For egg production, the energy demand of manure management is more than offset by the value of chicken manure as a fertilizer, hence this can have a net negative value for energy use (Table 2.2). There is less variation in the energy mix for livestock production. About a third comes from oil and a third from natural gas. 70–90 % of the energy utilized is for feed production and supply.

Energy Contribution to Food Processing

Post-producer

According to a recent USDA report processing industry energy use for cooking, cooling, and freezing contributes an average share of 15–20 % of total US food system energy use (Canning et al. 2010). The analyses used two US benchmark input-output accounts and a national energy data system to review energy usage and change in food production and consumption in the US over the decade from

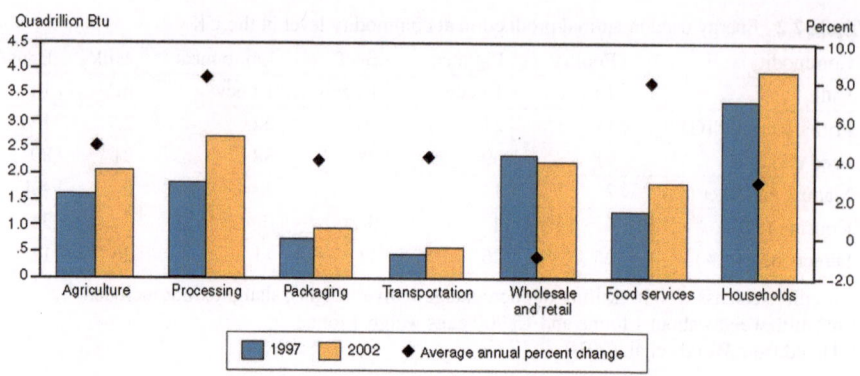

Fig. 2.1 Change in US energy consumption, by stage of production from 1997 to 2002. Adapted from Canning et al. (2010)

1997 to 2007 (Fig. 2.1). Half of the growth in food-related energy in the first five years of the decade was explained by a shift from human labour to energy services driven activities. Both households and food service industries outsourced manual food preparation and clean-up to food processing manufacturers, which typically used energy-based machinery to complete the tasks.

A time-use study of adults between ages 18 and 64 demonstrated that average time/day spent on cooking and cleaning at home was reduced from just over one hour in 1965 to half an hour in 1995 (Cutler et al. 2003). The decrease in food preparation time coincides with the growth in demand for convenience foods which require more processing, industrial preparation and packaging than food prepared at the household level. Food service establishments like restaurants and convention centres also increasingly outsource food preparation to the food processing industry. As a result energy flows through the food processing industries have increased in the US at an average rate of 8.3 % per annum between 1997 and 2002 (see Fig. 2.1). The USDA estimates that this flows through on a per capita basis to the equivalent of 24 gallons (90.8 L) of petroleum per person per year (Canning et al. 2010).

Consumer or Household/Kitchen Operations

The USDA report cited above also refers to the 2001 Residential Energy Consumption Surveys (RECS), used to obtain estimates of food-related household operational expenditures in 2002. According to these data, cooking (electric range, oven, microwave, toaster oven, and coffee makers) accounted for 6.5 %, refrigeration 14 %; freezing, 3.4 %; dishwashers, 2.5 % of household energy use. Combined, these sources accounted for 10 % energy use in the US food system and 26 % of the total proportion of household electricity in 2001.

Wholesale/retail activities account for about 4 % of the energy use in the US food system. These decreased by 1.1 % per annum from 1997 to 2001, probably driven by consolidation in the grocery sector.

Transportation

The concept of 'food miles' has become popular in the community as a way of assessing energy use for consumer purchasing decisions. However, some findings (Canning et al. 2010; Pelletier et al. 2011) suggest that energy flows associated with the commercial transportation of food represent less than 5 % of total energy use by the overall food system. This is largely because the bulk of food supplies are transported domestically in the US by road and rail and internationally by rail or ships, the latter being relatively energy efficient. Of course, this share of energy use is considerably higher for some food categories, such as fresh fruits and vegetables, and produce like fish, which require refrigeration or freezing and airfreight.

To maximize net energy savings through reliance on local food production, the local farm, agribusiness, and processing industries would need to be at least as energy efficient as the distant industry alternatives that they replace, whether produced domestically or in a foreign country.

References

Canning, P., A. Charles, S. Huang, K.R. Polenske, and A. Waters. 2010. Energy use in the US food system. *Economic research report number 94*. U. S. D. of Agriculture.

Cutler, D.M., E.A. Glaeser, and J.M. Shaprio. 2003. Why have Americans become more obese? *Journal of Economic Perspectives* 17(3): 93–118.

Pelletier, N., E. Audsley, S. Brodt, T. Garnett, P. Henriksson, A. Kendall, K.J. Kramer, D. Murphy, T. Nemecek, and M. Troell. 2011. Energy intensity of agriculture and food systems. *Annual Review of Environment and Resources* 36: 223–246.

Woods, J., A. Williams, J.K. Hughes, M. Black, and R. Murphy. 2010. Energy and the food system. *Philosophical Transactions of the Royal Society B: Biological Sciences* 365(1554): 2991–3006.

Chapter 3
Food, Water and Energy Nexus

Food, energy and water use are closely linked. Water production requires energy, energy production utilises water, and food production requires both water and energy! This complex interdependence was identified and explored in detail at the World Economic Forums in Davos from 2008 to 2013.

Interaction Between Food, Water and Energy Use

In 2011 70 % of the world's freshwater withdrawals were used for agriculture, and it took one litre of water to grow one calorie of food (WEF 2011). Water is also used intensively for the production of energy as this sector is the largest industrial user of water. The US Geological Survey estimates that to produce and burn the one billion tonnes of coal used per annum, mining and utilities withdraw between 208 and 284 trillion litres of water annually, i.e., 50 % of all the annual freshwater withdrawals in the US. Renewable energy also places a demand on water. Conventional biofuels like ethanol, utilize water for irrigation and solar thermal power plants utilize water for cooling. There is also a reverse link with energy required to access water, energy is needed to pump ground water and fresh flows as well as for treatment of potable and recycled water. In regions where fresh water is scarce, desalination of seawater to produce fresh water is utilized and this process is very energy intensive (Herndon 2013).

The forecast increases in demand for food, based on forecast population growth, increased urbanization and changes to diet as a result of an increasing global middle class, in turn increase the demand for energy and both increase the demand for water (WEF 2011; Bizikova et al. 2013). In addition, the impact of climate change projections need to be taken into account when examining the global capacity to sustainably meet the forecast demand for food production for the rest

© The Author(s) 2015
M. Verma, *Energy Use in Global Food Production*,
SpringerBriefs in Food, Health, and Nutrition, DOI 10.1007/978-3-319-16781-7_3

of this century (IPCC 2014). Both these factors are examined in more detail below, followed by a possible approach to planning for long term sustainability in the water, energy and food nexus (WEF).

Global Population Growth Forecasts

The United Nations 2012 Revision report (United Nations 2013) is a review of past global demographic trends and future prospects and provides a comprehensive basis for estimates and projections of global population out to the end of this century. The report models three scenarios, low-variant, medium-variant and high variant, based on projected changes in average total fertility. According to the medium-variant projection global population will increase over the next decade, reaching 8.1 billion by 2025, further increase to 9.6 billion in 2050 and stabilize at about 10 billion by 2100 (Fig. 3.1). The medium-variant projection assumes a decline of fertility for countries where large families are still prevalent and a slight increase in fertility for countries with fewer than two children per woman on average.

It is worth noting that small differences in fertility over the next decade will have a major impact on the final global population size in 2100. For example, compared to the medium-variant, the high-variant projection assumes an extra half a child per woman on average, resulting in 16.6 billion population by 2100. The low-variant projection, by contrast, assumes half a child less per woman on average, producing a peak population of 8.3 billion by 2050.

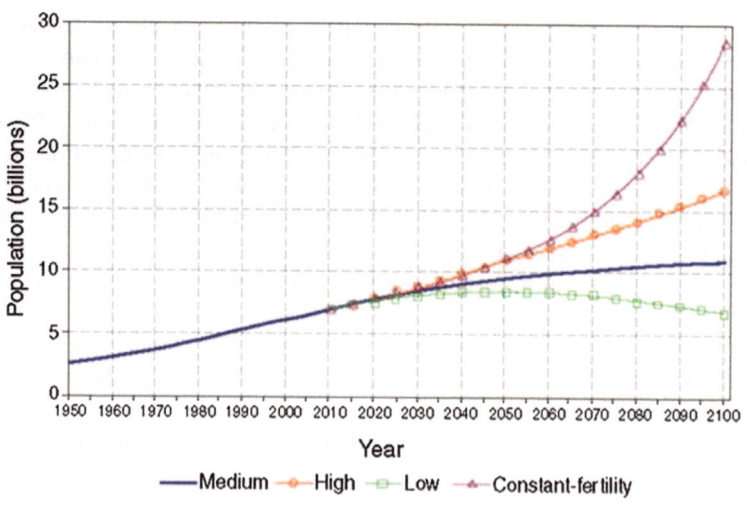

Fig. 3.1 Population of the World; forecast scenarios. Adapted from (United Nations 2013)

The regional distribution of these projected increases in population is also important; almost all of the additional 3.7 billion people will enlarge the population of the developing countries.

Traditionally the world has been segmented into the 'developed' and 'developing' world. The former are developed, post-industrial societies that have gone through a demographic transition such that they are characterized by stable or declining populations which are increasing in median age. Most population growth in these societies comes from net inward migration. The latter group can now be said to consist of two populations: the 'late-stage developing' nations, which are industrializing rapidly and where population growth rates are decelerating, with an increase in affluence and age profile; and 'newly developing' nations that are beginning to industrialize, with high to very high population growth rates and a predominantly young age profile.

The projected trends in the 2012 Revision are contingent on fertility declines in late-stage and newly developing countries. Fertility in less developed regions as a whole is expected to drop from 2.69 children per woman in 2005–2010 to 2.29 by 2050 and 1.99 by 2100. In the 49 least developed countries, the projected reduction is even steeper from 4.5 children per woman to 2.9 children by 2050 and to 2.11 by 2100. To achieve such reductions it is essential that access to family planning is expanded, particularly in the least developed countries. In 2013, it was estimated that just 31 % of women of reproductive age in a relationship, used modern contraceptive methods, with a further 23 % of such women experiencing an unmet need for family planning.

The median age, that is, the age that divides the population in two halves of equal size, is an indicator of population ageing. Globally, the median age is projected to increase from 29 to 36 years between 2013 and 2050 and to 41 years by 2100. The median age is higher in countries or regions that have been experiencing low fertility for a long period of time. Europe today has the oldest population, with a median age of 41 years in 2013. This is expected to reach 46 years by 2050 and then 47 years by 2100.

If the world population stabilizes at 10 billion at the end of the century, as projected under the median-variant scenario, an additional 3 billion people will require food, water and energy.

Impact of Increasing Urbanization and Rising Affluence on Water, Food and Energy

"The global rise of cities has been unprecedented. In 1800, 2 % of the world's population lived in cities. Now it's 50 %. Every week, some 1.5 million people join the urban population, through a combination of migration and childbirth." says Ian Powell, Senior Partner at PwC commenting on the 2014 Global Report (Powell 2014).

Globally more people live in cities than in rural areas (United Nations 2014). In 2007, for the first time in history, the global urban population exceeded the global rural population. The world has experienced rapid urbanization over the last six decades. In 1950, 70 % of people world-wide lived in rural settlements and fewer than 30 % in urban areas. In 2014, 54 % of the world's population is urban. and the expectation is that by 2050, more than 66 % of the population will live in urban areas.

While at present there remains a great diversity in the characteristics of the World's urban environs (United Nations 2014), with about half of urban dwellers living in relatively small settlements of fewer than 500,000 inhabitants each, by 2015, it is projected that about 600 million people will live in megacities, with greater than 5 million people in each (Kraas 2007). By 2030 the UN projects that there will be 41 urban agglomerations housing 10 million people each, with some of the fastest growing megacities occurring in the developing world.

Megacities require huge natural and human resources for energy, water, food, industry, infrastructure and services (Kraas 2007) and by one estimate, although cities cover just 0.5 % of the earth's surface they consume 75 % of its resources (Powell 2014).

Projections of population growth drive forecasts that suggest an additional 60–70 % demand for agricultural products by 2050. Rising affluence and an increase in the middle class sub-population in the developing world also suggests a shift away from predominantly grain-based diets to consumption of more meat and animal products (Foresight 2011). This has implications for water use as well. Since, producing 1 kg of rice, for example, requires about 3,500 L of water, 1 kg of beef some 15,000 L, and a cup of coffee about 140 L, this dietary shift has had the greatest impact on water consumption over the past 30 years.

Potential Impact of Climate Change on Food Production

The Intergovermental Panel on Climate Change in their recent Approved Summary for Policy Makers (IPCC 2014) make the statement that anthropogenic climate change (i.e., the change caused by human activities and actions) is impacting global hydrological systems, affecting water resources in terms of both quantity and quality. Notably, negative impacts of climate change on crop yields in many regions outweigh positive impacts of increased seasonal rainfall.

Indeed, it is likely that climate change has already more than doubled the probability of the occurrence of heat waves in some of the regions that are currently net exporters of food staples (FAO 2013; IPCC 2014).

Future scenarios and projections of climate change are based on forecast cumulative emissions of CO_2-eq. Anthropogenic greenhouse gas (GHG) emissions are mainly driven by population size, economic activity, lifestyle, energy use, land-use patterns, technology and climate policy. IPCC uses "Representative Concentration pathways" (RCPs) for making projections on these factors for four different 21st century scenarios. The stringent mitigation scenario (RCP2.6) aims to keep average global warming below 2 °C above pre-industrial (1861–1880) temperatures. 'Business as usual' scenarios with minimal efforts to curb or constrain emissions are described in RCP6.0 and in RCP8.5 (see Fig. 3.2).

Multi-model results show that limiting total human-induced warming under RCP2.6, with a probability of >66 % would require cumulative CO_2 emissions from all anthropogenic sources since 1870 to remain below about 2900 $GtCO_2$ (*with a range of* 2550–3150 *$GtCO_2$ depending on non-CO_2 drivers*). About 1900 $GtCO_2$ had already been emitted by 2011.

IPCC 2014 states that "*The increase of global mean surface temperature by the end of the 21st century (2081–2100) relative to 1986–2005 is likely to be 0.3–1.7 °C under RCP2.6, 1.1–2.6 °C under RCP4.5, 1.4–3.1 °C under RCP6.0, and 2.6–4.8 °C under RCP8.5*", and that "*Changes in precipitation will not be uniform. The high-latitudes and the equatorial Pacific are likely to experience an increase in annual mean precipitation under the RCP8.5 scenario. In many mid-latitude and subtropical dry regions, mean precipitation will likely decrease, while in many mid-latitude wet regions, mean precipitation will likely increase under the RCP8.5 scenario. Extreme precipitation events over most of the mid-latitude land masses and over wet tropical regions will very likely become more intense and more frequent.*"

Ocean warming over the last few decades accounts for 90 % of the energy accumulated in the climate system between 1971 and 2010, with most of the warming occurring near the ocean surface, i.e., the upper 75 m (IPCC 2014). In addition the uptake of CO_2 by the oceans has resulted in acidification. The pH of the ocean surface water has already decreased by 0.1 units, which corresponds to a 26 % increase in acidity since the beginning of the industrial era (IPCC 2014). Acidity is projected to increase significantly by the end of the century under the IPCC scenarios (see Fig. 3.2).

The net result of these changes is that climate change will impact food security. Global marine species redistribution and reduction of marine biodiversity in some regions will challenge fisheries productivity.

For staple crops like wheat, rice and maize in tropical and temperate regions a > 2 °C rise will have a negative impact on production, though some individual regions may benefit. Climate change is also projected to reduce renewable surface water and ground water resources in most dry subtropical regions, intensifying competition for water between sectors.

Specific examples of impact of warmer temperatures on crop yields come from various studies around the globe.

The US produces 41 % of the world's corn and 38 % of the world's soybeans (Schlenker and Roberts 2009). The authors found that temperature increases above a threshold value for these crops was very harmful. Average yields for these crops are predicted to decrease by 30–46 % before the end of the century under the slowest warming scenario.

A study by Kansas State University scientists reviewed 55 years' worth of historical wheat yield data from western Kansas (Southwest Farm Press 2011).

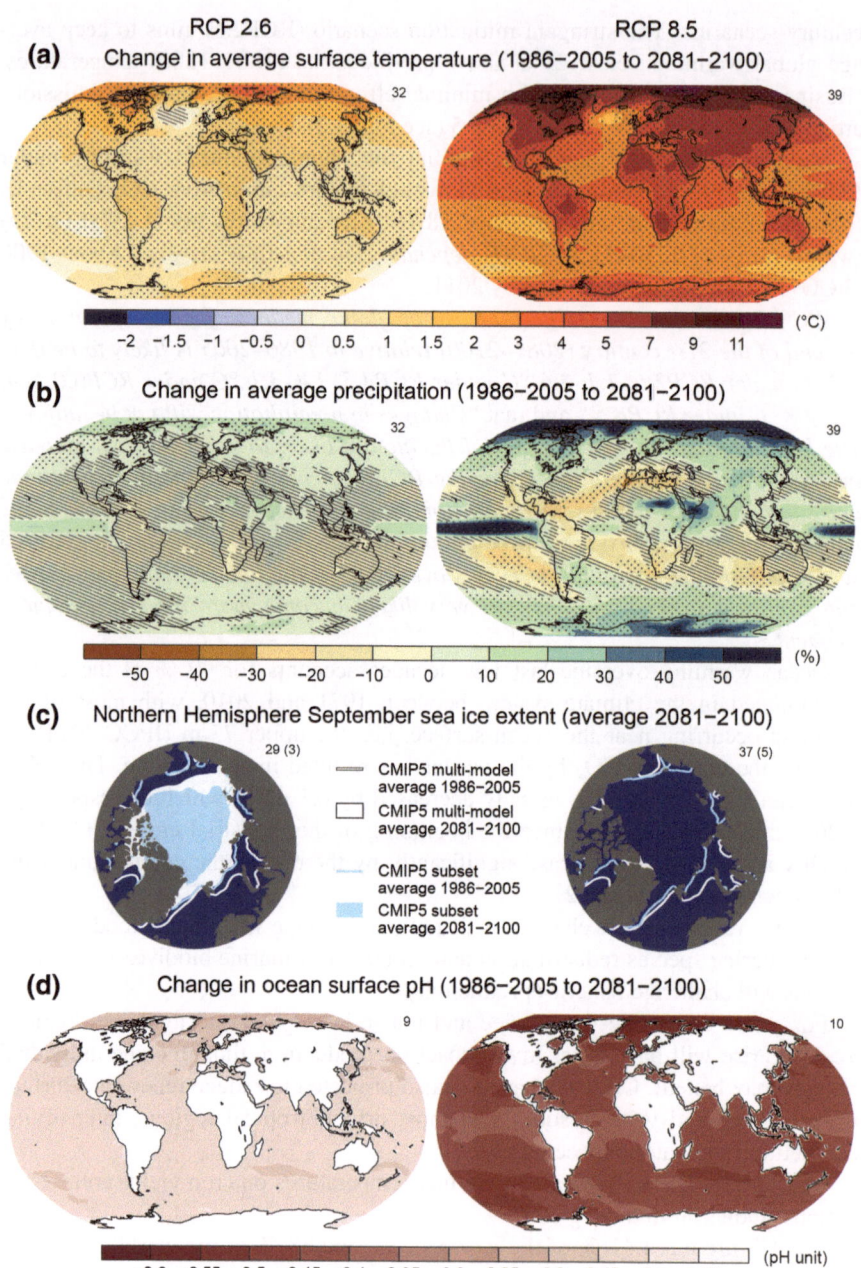

Fig. 3.2 Maps of projected late 21st century annual mean surface temperature change, annual mean precipitation change, Northern Hemisphere September sea ice extent, and change in ocean surface pH. **a** Change in average surface temperature (1986–2005 to 2081–2100). **b** Change in average precipitation (1986–2005 to 2081–2100). **c** Northern hemisphere September sea ice extent (average 2081–2100). **d** Change in ocean surface pH (1986–2005 to 2081–2100). Reproduced from Figure SPM.8 (IPCC 2013)

For every 1 °F increase in May, irrigated wheat yields increased 1 bushel per acre. However, for a 1 °F increase in June, October or November, wheat yields decreased about 1 bushel per acre.

In India, various reports summarized the impact and vulnerability of Indian agriculture to climate change, (Aggarwal et al. 2010). Simulation studies indicate a possibility of loss of 4–5 million tons in wheat production with every rise of 1 °C temperature throughout the growing period even after considering carbon fertilization. Rice production in Tamil Nadu could be reduced by 30–35 % by 2050 due to forecast changes in temperature and rainfall. Groundnut yields are projected to decline by 7 % in the medium term. Milk production and various fisheries are also forecast to show reduced productivity depending on the region.

Sustainability of the Food, Water and Energy Nexus

Water is an essential ingredient for life, and as illustrated above is required for both food and energy security. However, the pressures on water supplies from a growing global population and increasing global affluence are currently relentless. Global demand for food and energy are likely to more than double by 2050, thereby doubling the demand for water. Yet at the moment aquifers and rivers are overused and under unprecedented pressure. The current drought in California provides an example of the unsustainable use of aquifers for both agriculture and energy generation (fracking) (Dimick 2014).

There are however, new approaches to water management that are paying dividends. For example, the recent millennial drought in Australia led to a mix of technology, policy and lifestyle changes which allowed the community to adapt to a 70 % decrease in water availability in Australia's largest irrigated agricultural region, the Murray-Darling Basin, without negatively affecting GDP (Maywald 2013). In sub-Saharan Africa where 95 % of crops are rain-fed, only 10–30 % of available rainfall is being used productively; rain water harvesting, better terracing and tiling could provide large benefits to water usage (Editorial 2008). Thus, low tech solutions can make a difference to the rate of water usage in various parts of the world. However, for a truly sustainable approach to be developed worldwide, coordinated policy actions are required across countries and regions.

Fortunately, the approach taken at recent Davos World Economic Forums has commenced laying out a framework for policy development (Bizikova et al. 2013). Bizikova and coworkers set out practical applications for integrated management of the Water Food and Energy (WEF) nexus at local, regional and national levels. The growing recognition of relationships among the elements of WEF has highlighted the need to assess impact and consequences on all three elements when dealing with global challenges in this domain.

The WEF Security Nexus can be summarized as follows:

Food security

Food availability: production, distribution and exchange of food
Access to food: affordability, allocation and preference
Utilization: nutritional value, social value and food safety
Food stability over time

Water security

Water access, water safety and *water affordability* so that every person can lead a clean, healthy and productive life, while ensuring that the natural environment is protected and enhanced.

Energy security

Continuity of energy supplies relative to *demand*
Physical availability of supplies
Supply sufficient to satisfy demand at a given price

Bizikova and coworkers reviewed three WEF Frameworks for action using the new nexus-oriented approach. (i) The Bonn2011 Nexus Conference which recommended, among other things, using waste as a resource in multi-use systems; (ii) World Economic Forum, 2011 which outlined a number of areas to explore as levers including market-led resource pricing for managing the nexus; and (iii) International Centre for Integrated Mountain Development, 2012 focused on the Himalayas and South Asia, and was particularly aimed at restoration of natural water storage mechanisms. The ultimate focus of these frameworks is to promote action by providing policy entry points to reduce trade-offs, explore synergies and promote the transition to a more sustainable future.

Based on the existing frameworks, the International Institute for Sustainable Development (IISD) identified areas for intervention in promoting WEF (Bizikova et al. 2013). The areas are—Engaging Stakeholders; Improving Policy Development, Coordination and Harmonization; Governance, and Integrated and Multi-stakeholder Resource Planning; Promoting Innovation; and Influencing Policies on trade and investment in environment/climate.

IISD Water-Energy-Food Security Analysis Framework was developed as a result of the work by Bizikova and coworkers, to bridge some of the gaps in the existing frameworks. The IISD approach is centered on an integrated approach to ecosystem management.

Specifically, they recognize that ecosystems provide the goods and services (EGS) that humans rely on, they provide water, food and energy and influence their supply, availability and access. Restoring and managing EGS provides a practical way to optimize WEF security. This framework is place-based, with a geospatially explicit, ecosystem-based approach. The goal of the framework is to inform investment, decision-making and risk management to ensure optimization of WEF security.

The fundamental premise is that ecosystem goods and services are benefits we receive from well-functioning ecosystems. Also that a temporal perspective is important to avoid trading off security today for security tomorrow—i.e., pushing externalities to the future.

Implementation of the framework requires a series of analyses:

1. Build three independent security frameworks, one around utilization of each of the WEF elements.
2. Identify access, i.e., how watershed communities access their water, food and energy, including water flows, agriculture, food production, energy supply in the context of natural systems and human systems.
3. Explore the different combinations of elements and aspects of use, access and availability to clarify the relationships and draw attention to particular combinations of elements.
4. Assess the impact of the natural and human built systems on the access and stable supply of WEF. For example, natural or constructed wetlands can influence both availability and quality of water, or infrastructure that can provide access routes for food transportation can also be used for water supply and energy grids. A third example is waste management specific to food or water, which could benefit all three elements.
5. Overlay governance, management systems, markets, and existing policies.

IISD developed a graphical model (Fig. 3.3a, b) for their framework to provide analysts and decision-makers with a menu approach to watersheds and communities to assist with identification and development of priorities and risks to optimize WEF security.

The next steps involve a practical participatory planning process with four main stages:

Stage 1: Assessing the Water–Energy–Food Security System
Stage 2: Envisioning Future Landscape Scenarios
Stage 3: Investing in a Water–Energy–Food Secure Future
Stage 4: Transforming the System

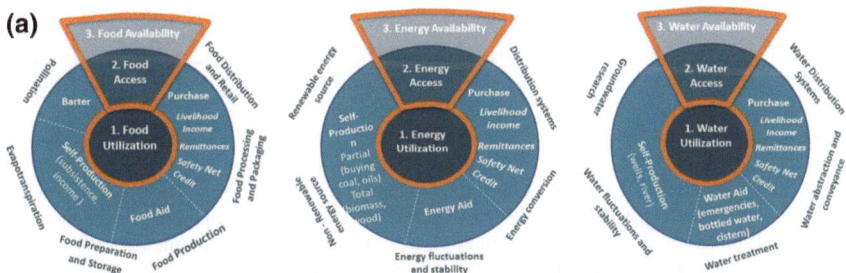

Fig. 3.3 **a** Graphical overview of the framework linking water, food and energy security—defining key securities as core elements of WEF (analysis 1–3). **b** Graphical overview of the framework linking water, food and energy security—overlaying natural and built systems and governance considerations (adding analyses 4 and 5). Image republished with the permission of the International Institute for Sustainable Development (IISD)

(b)

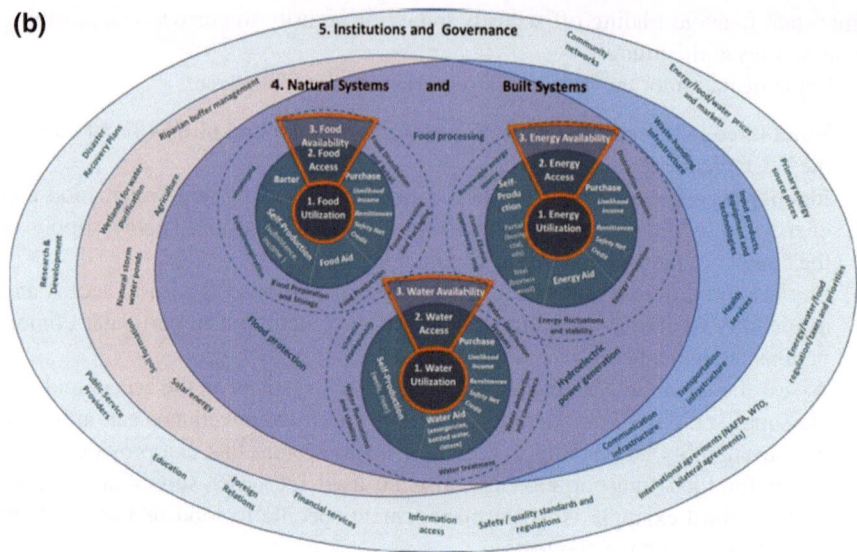

Fig. 3.3   (continued)

To transition towards sustainability and WEF security, the proposed IISD Framework (Bizikova et al. 2013) emphasizes communication and cooperation at all levels and a focus on implementation to minimize the 'implementation gap' that besets the number of strategies developed in this area.

For this approach to be truly sustainable, a global change in thinking is required with a definition of community that involves multiple nation states and regions, followed by meaningful cooperation across the community.

References

Aggarwal, P.K., S. Kumar, S.N., and H. Pathak. 2010. *Impacts of climate change on growth and yield of rice and wheat in the Upper Ganga Basin*. Pusa: Indian Agricultural Research Institute (IARI).

Bizikova, L., D. Roy, D. Swanson, H.D. Venema, and M. McCandless 2013. *The water–energy–food security nexus: towards a practical planning and decision-support framework for landscape investment and risk management*. Canada: International Institute for Sustainable Development.

Dimick, D. 2014. If you think the water crisis can't get worse, wait until the aquifers are drained. Retrieved 30 Dec 2014, from http://news.nationalgeographic.com.au/news/2014/08/140819-groundwater-california-drought-aquifers-hidden-crisis/.

Editorial, N. 2008. A fresh approach to water. *Nature* 452(7185): 253.

FAO. 2013. *FAO stastical yearbook—world food and agriculture*. Rome: Food and Agriculture Organisation of the United Nations.

Foresight. 2011. *The future of food and farming: challenges and choices for global sustainability*. UK Government Office for Science.

Herndon, A. 2013. Energy makes up half of desalination plant costs: study. From http://www.blo omberg.com/news/2013-05-01/energy-makes-up-half-of-desalination-plant-costs-study.html.

IPCC. 2013. Summary for policymakers. In Climate change 2013: the physical science basis. Contribution of working group I to the fifth assessment report of the intergovernmental panel on climate change, ed. T.F. Stocker, D. Qin, G.-K. Plattner, M. Tignor, S.K. Allen, J. Boschung, A. Nauels, Y. Xia, V. Bex and P.M. Midgley, 1–30, Cambridge.

IPCC. 2014. Climate Change 2014 Synthesis Report.

Kraas, F. 2007. Megacities and global change: key priorities. *Geographical Journal* 173(1): 79–82.

Maywald, K. 2013. Water 2013 and beyond—sustaining the momentum. Perth: OzWater 13.

Powell, I. (2014). Global annual review. Retrieved 27 Dec 2014, from http://www.pwc.com/gx/en/issues/megatrends/rapid-urbanisation-ian-powell.jhtml.

Schlenker, W., and M.J. Roberts. 2009. Nonlinear temperature effects indicate severe damages to U.S. crop yields under climate change. *Proceedings of the National Academy of Sciences* 106(37): 15594–15598.

Southwest Farm Press. 2011. 55-year Kansas wheat study examines yield effects of precipitation, temperature and timing out west, from http://southwestfarmpress.com/grains/55-year-kansas-wheat-study-examines-yield-effects-precipitation-temperature-and-timing-out-we?page=1.

United Nations, D. 2013. *World population prospects: the 2012 revision, volume I: comprehensive tables ST/ESA/SER.A/336.* New York: Department of Economic and Social Affairs, Population.

United Nations. 2014. *World urbanisation prospects, 2014 revision.* New York: United Nations.

WEF, W.E.F. 2011. *Water security-the water-food-energy-climate nexus.* Washington: World Economic Forum.

Chapter 4
Food Wastage—Energy Wasted

Global Extent of Food Wastage

According to the UN (2014), **roughly 30 % of the food produced worldwide—**
about 1.3 billion tons—**is lost or wasted every year**, which means that the water
and energy used to produce it, is also wasted. Poor storage facilities, over-strict
sell-by dates, "get-one-free" offers, and consumer fussiness all contribute to the
waste, according to the UK's Institution of Mechanical Engineers (IME 2013).

In low-income, developing countries, most food loss is during production,
whereas in higher-income developed countries the food waste is at the consumption stage. This food wastage is unsustainable in the face of an increasing global
population.

Overall, on a per-capita basis, much more food is wasted in the industrialized
world than in developing countries. It was estimated that the per capita food waste
by consumers in Europe and North-America is 95–115 kg/year, while this figure
in sub-Saharan Africa and South/Southeast Asia is only 6–11 kg/year (FAO 2011).

There is potential to provide >40 % more food by simply reducing or eliminating
food losses and wastage.[1] This would also reduce the pressure on land, water and
energy resource to feed the growing global population.

---

[1] FAO defines terms as follows: *"Food losses refer to decreases in edible food mass throughout
the part of the supply chain that specifically leads to edible food for human consumption. Food
losses take place at production, post-harvest and processing stages in the food supply chain.
Food losses occurring at the end of the food chain (retail and final consumption) are rather
called "food waste", which relates to retailers' and consumers' behavior."*

© The Author(s) 2015
M. Verma, *Energy Use in Global Food Production*,
SpringerBriefs in Food, Health, and Nutrition, DOI 10.1007/978-3-319-16781-7_4

Regional Variation; Cause and Effect

Globally about four billion tonnes of food are produced per annum (FAO 2013). That amounts to about 1.5 kg/capita per day on average across the globe. However, production, consumption and wastage are not evenly distributed around the world.

In newly developing regions of the world, food losses tend to occur primarily at the farmer-producer end of the supply chain due to inefficient harvesting, inadequate transportation and poor storage infrastructure. In late-stage developing regions the food loss moves up the supply chain due to deficiencies in transportation and associated infrastructure. In developed parts of the world food loss and waste is driven largely by modern consumer culture and retail and customer behaviour. A larger proportion of the food produced reaches markets and consumers, but entire crops of fruit and vegetables may be rejected by retailers because the physical appearance does not meet the marketing standard and consumer expectation. Of the food that does reach retail, 30–50 % of purchased food is thrown away by the consumer (IME 2013).

Post—Harvest Waste

In the newly developing regions of the world, such as sub-Saharan Africa and South East Asia, wastage at the farmer-producer end of the supply chain is largely due to poor handling and storage of crops. As a result bruising of produce, fungal infestations and rodents destroy or degrade large quantities of foodstuffs. In South-East Asian countries for example, rice losses can range from 15 to 80 % of the entire production, amounting to a total of 180 million tonnes per annum at the higher end. However, in a late-stage developing country like China, the loss is much lower (Mejía 2004; IME 2013).

In developed countries, more of the produced crop reaches the wholesale area due to better infrastructure, including transport, storage and processing facilities. However, a portion of the produce does not reach the retailer or consumer.

Wastage rates are higher for vegetables and fruit than for grains. Recent reports from studies and surveys from UK and India illustrate the level of wastage in each region (IME 2013). At a World Potato Congress in 2012, it was reported (Bowen 2012) that in 2008 up to 45 % of potatoes grown in the UK were not delivered to the retail market; 6 % was lost in the field, 12 % discarded on initial sorting, 5 % lost in storage, 1 % was lost on post-storage inspection and 22 % was lost due to rejection after washing. While in India, the Institution of Mechanical Engineers reports (IME 2013) that 40 % of ALL fruit and vegetable is lost between the grower and consumer due to lack of refrigerated transport, poor roads, inclement weather and official corruption.

Field wastage in developed countries is driven by the supply contracts, where penalties are imposed on growers for failure to deliver contracted quantities. As a result farmers often cultivate more crop than they require, as a form of insurance against uncontrollable variables. As a result of these factors, in the UK it is estimated that 30 % of the vegetable crop is never harvested. In less developed countries, most agricultural operations are conducted by manual labour thus, harvesting can involve multiple handling of produce, from field to farmyard, to on- or off-site storage and transport to point of sale. Produce is lost or damaged all along this chain, by bruising or spillage and contamination.

Wastage during storage has different drivers, since most food crops are harvested annually and must be stored to provide an adequate supply throughout the year or to create a buffer between supply and demand. Under ideal conditions cereals like wheat and maize can be stored for up to five years. However, perishables like fresh fruit and vegetables, eggs, meat and dairy need very closely controlled conditions for storage for any length of time. Grain wastage in storage is reported (IME 2013) to vary widely across regions, from 0.75 % of stored grain in a developed country like Australia to an annual loss of up to 21 million tonnes of wheat in India, or 10 % of production, due to inadequate storage and distribution (GDF 2011).

Waste in transportation is generally caused by inadequate handing facilities. Losses can be minimized by the introduction of appropriately designed modular packing crates along with the infrastructure to transport and store the crates efficiently.

Retail and Post—Consumer Waste

At both the market and retail output level there is also a considerable difference in wastage between developed and developing regions. For example, the logistics systems and infrastructure of modern supermarkets in urban centres around the globe ensures that perishable produce waste is minimized in the store. In developing regions on the other hand, wastage in open stalls and in market places is much higher (FAO 2011).

When it comes to domestic wastage by consumers, the picture changes drastically. It is estimated that consumers in the developed world waste 222 million tonnes of food per annum, which is almost equivalent to the entire net food production in sub-Saharan Africa (FAO 2011). Of the produce that reaches supermarket shelves, it was reported that in the UK in 2007, 22 % of food and drink purchased, or 8.3 million tonnes, was thrown away by the consumer post-purchase (WRAP 2013). Almost half of this was due to the food going off or not being used by the 'use by' or 'best before' dates. This food waste comes at a cost to the consumer's budget, but it is reported that the average household in the UK spends only 11 % of its budget on food (IME 2013), which may explain why it is not valued more highly. In the US estimates of food waste vary from 27 % in 1995 (Cuellar and Webber 2010) to almost 50 % of food produced being wasted (Bloom 2010).

In developing regions, patterns of food waste vary between rural and urban residents. In rural areas food is typically stored from harvest of staple crops and is subject to spoilage by rodents and pests due to inadequate and primitive storage at the household level. In urban areas, wastage is reduced to an absolute minimum by the process of purchasing just the amount required for the day or a minimum number of meals.

Food wastage has a negative multiplier effect as it equates to wastage of the embedded energy in the food, and wastage of the water used to produce it. Therefore, reducing food wastage at every level is an important consideration for improving future food security in a sustainable manner.

References

Bloom, J. 2010. *American wasteland—how America throws away nearly half of its food (and what we can do about it)*. New York: Perseus Books Group.

Bowen, S. 2012. War on waste in the potato supply chain. http://www.potato.org.uk/news/war-waste-potato-supply-chain.

Cuellar, A.D., and M. Webber. 2010. Wasted food, wasted energy: The embedded energy in food waste in the United States. *Environmental Science and Technology* 44: 6464–6469.

FAO. 2011. *Global food losses and food waste—extent, causes and prevention*. Rome: Food and Agriculture Organisation of the United Nations.

FAO. 2013. *FAO statistical yearbook—world food and agriculture*. Rome: Food and Agriculture Organisation of the United Nations.

GDF. 2011. Grain depot fund prospectus. http://sustainableinvestingchallenge.org/b6.pdf

IME. 2013. *Global food; waste not, want Not*. London: Institution of Mechanical Engineering.

Mejía, D.J. 2004. *Rice post-harvest system: an efficient approach*. Rome: Food and Agriculture Organisation of the United Nations.

United Nations, D. 2014. *World urbanisation prospects, 2014 revision*. New York: United Nations.

WRAP. 2013. Household food and drink waste in the United Kingdom 2012.

Chapter 5
Energy Intensity and Efficiency in Food Production

The amount of food produced globally increased dramatically over the latter part of the 20th century, for a variety of reasons.

Productivity growth for the production of grains was high from the 1960s to the 1980s due largely to the 'Green Revolution'. This increase in productivity was attributable to higher yields per hectare rather than increases in cropping area and was achieved by significant increases in use of fossil fuel for farm equipment, machinery, irrigation and chemical fertilizer and pesticide inputs (Rayner et al. 2011). In the U.S. farmers increased yields of corn, on average, to 134 bushels per acre in 1999 from 30 bushels per acre in 1920, an increase of almost 350 % (Horrigan et al. 2002). At least one-third of crop yield increases can be attributed to the application of mineral fertilizers. In addition, irrigated agriculture (which uses energy to deliver water) accounts for 40 % of the world's food production (FAO 2012a). These increases have been achieved while globally land available for agriculture has decreased per capita from 1.3 to 0.72 ha from 1967 to 2007 (Leaver 2011). Therefore, increases in food production have been achieved by increasing the overall *energy intensity* of food.

Forecast increases in food demand due to population increase and changes in food consumption patterns will require 60–70 % more food to be produced by 2050 (Leaver 2011; FAO 2012b). If this increased demand is met by simple extrapolation of the current methodologies, an additional 3.43 Gha of land would be required, or a total of 8.33 Gha which is 83 % of total productive land on the planet. Clearly, this would have a severe negative impact on species diversity and quality of life for future generations of humans.

An alternative approach is required to achieve the expected necessary increase in food availability for global food security, one that uses less non-renewable energy per hectare of land and per unit of food output. This will require innovations in renewable sources of energy, more efficient methods of application of inputs to crops and alternative means of producing crop fertilizers and animal feeds.

© The Author(s) 2015
M. Verma, *Energy Use in Global Food Production*,
SpringerBriefs in Food, Health, and Nutrition, DOI 10.1007/978-3-319-16781-7_5

FAO's Energy-Smart Food Production Programme

The key message from FAO is "*Making a gradual shift to energy-efficient agrifood systems that make greater use of renewable energy technologies and better integrate food and energy production, may be the most viable solution for simultaneously reducing agrifood systems' dependency on fossil fuels and building their resilience against higher energy prices. This shift to energy-smart agrifood systems can also improve productivity in the food sector, reduce energy-poverty in rural areas and contribute to achieving goals related to national food security, climate change and sustainable development.*" (FAO 2012a).

To this end, FAO launched the Energy-Smart Food for People and Climate (ESF) Programme in 2012. This multi-partner initiative is designed to assist member countries make the shift to more sustainable food production systems, through increased energy efficiency, diversification and access.

Energy Intensity of Food Production

FAO's Programme on Sustainable Crop Production Intensification (SCPI) was established with the recognition that a paradigm shift is required to sustainable intensification of crop production to ensure future food security. SCPI aims to find intensification solutions through an ecosystem approach with appropriate external inputs applied in the right amounts. Farming systems for sustainable crop production intensification are based on three technical principles (FAO 2012a):

i.   simultaneous achievement of increased agricultural productivity and enhancement of natural capital and ecosystems services;
ii. higher rates of efficiency in the use of key inputs, including water, nutrients, pesticides, energy, land and labour; and
iii. the use of managed and natural biodiversity to build system resilience to abiotic, biotic and economic stresses

The Programme promotes 'conservation agriculture' with reduced tillage, judicious use of organic and inorganic fertilizer, integrated pest and weed management and efficient water management. Use of these methods should contribute to reducing the use of fossil fuel.

No-till crop production is a promising way to reduce energy input into agriculture and zero tillage has been promoted by FAO for many years as a way of lowering fossil fuel consumption and improving yields at the smallholder level. It was reported (Pelletier et al. 2011) that energy consumption for conventional tillage ranges from 412 to 740 MJ/ha. By contrast, energy consumption for mulched systems ranged from 183 to 266 MJ/ha, compared to 80–284 MJ/ha for no-till. In the U.S. the land under conservation agriculture, which includes no-till methods, increased from <15 % in 1980 to about 35 % by 1993 (Horrigan et al. 2002). No-till farming is gaining strong traction in other developed countries as well. Chapter 6 reports on a case-study from Australia.

Inefficiency in fertilizer use leads to both economic losses and environmental damage and the SCPI programme promotes better management of fertilizers, including method and timing of application as well as crop rotation to optimize use of mineral fertilizers. China's Lake Taihu is a well-known example of a large body of water that has deteriorated over the last two decades due to excessive nutrient run-off, in part as a result of overuse on farms (Stone 2011). Lake Eyrie in the Great Lakes region of North America, suffered from toxic algae blooms in the 1970s from excessive fertiliser use and run-off from neighbouring farms and sewage-treatment works. Major efforts to reduce phosphorus loading were successful and prevented algal blooms till the massive toxic algae bloom seen in 2011. This time, severe spring rains washed away farm-applied nutrients increasing the load and contributing to the problem. Improved forecasting of spring storms to better guide the timing of fertilizer application may help to prevent this loss of nutrients and consequential environmental damage (Borre 2013).

Integrated pest management methods are also promoted to reduce the overall use of pesticides and their associated 'embodied' fossil fuel energy.

An associated focus is on developing appropriate mechanization in the form of correctly sized and designed sowing and harvesting equipment, for conservation agriculture in the developing world, again reducing the amount of fossil fuel used on farm.

Apart from crop production for grains, fruit and vegetables, the food chain includes animal-derived products from fish and other seafood, as well as from livestock production.

Fisheries are an important component of the global food supply, providing 16 % of all protein consumed and 6 % of all protein consumed by people. Capture fishing has traditionally been the primary source of supply. Today however, aquaculture contributes almost half of the fish supply for human food and this is set to expand. FAO's Fisheries and Aquaculture Department developed the first standardized surveys of the fishing fleet in the 1980s. FAO reports that there are currently about 4.3 million vessels in the global fishing fleet and that capture fishing has become highly energy-use intensive, with fuel costs typically representing 30–50 % of operating costs (FAO 2012a). Again opportunities for fuel savings vary widely depending on the fleet, fishing conditions and market management. The use of energy in aquaculture is more indirect, chiefly relating to the procurement and processing of food for the fish. Growth in the aquaculture sector will rely on improving feeding efficiency, increasing the land or water-based productivity and finding replacements for fish-food currently sourced by capture fishing (Greiff 2015).

Food from livestock products account for about 13 % of all calories consumed by humans and provides 25 % of dietary protein. Meat consumption is projected to rise by >70 % and dairy consumption will grow about 58 % by 2050. It is difficult to envisage current energy-intensive animal-rearing operations expanding in a sustainable manner to keep pace with this demand. Energy costs of producing, processing and transporting animal feed are high, particularly for animals raised in intensive systems. In addition, intensive livestock production typically has a higher energy use than crop production per calorie of food produced. This is due to the inherent inefficiency in biological feed conversion by animals. Energy associated with feed provision for intensive livestock production is a large contributor to energy use. This is estimated to be 75 % of the energy used for livestock production in the UK and 86 % for feedlot beef in the U.S. (Pelletier et al. 2011).

Sourcing energy-efficient feeds and improving feed conversion efficiencies offers opportunities to decrease energy use. Better understanding of nutritional requirements and provision of better formulated feeds can improve livestock productivity and reduce negative environmental impacts. FAO reported that the shift from an unbalanced diet to a balanced diet for livestock on smallholder farms, reduced the amount of feed required and improved feed use efficiency by more than 30 %, simultaneously resulting in 15 % less methane production from ruminants (FAO 2012a). Livestock production can offer energy-smart solutions for meeting energy demands by using animal manure to produce both organic fertilizers and energy from biogas. FAO estimates that in developed countries, only around 15 % of nitrogen applied to soils comes from livestock manure. Better manure management strategies can be used to produce both organic fertilizers and renewable energy. Manure can be anaerobically digested to produce methane that can be captured and used for co-generation of energy. A study from piggeries in the EU showed that methane produced from manure could reduce grid-supplied energy use by 57 % (Pelletier et al. 2011). In the developing world, effluent from anaerobic digesters can also be used as a replacement for chemical fertilizers.

Energy Efficient Irrigation

More intensive use of irrigation was a major driver of increased productivity behind the green revolution. Globally, about 300 million hectares or 20 % of farmland is irrigated. This accounts for 70 % of all freshwater withdrawals and contributes 40 % of the world's food production (FAO 2012a). It is estimated that 40 % of water withdrawn is 'lost' by leakage or run-off. Distribution and storing water in irrigation systems increases the energy intensity per hectare of cropped land. Therefore, improving the efficiency of water use is critical to future food security.

Implementing effective solutions for saving energy and water is complex as there are tradeoffs. For instance, some water efficient systems like drip irrigation require more energy. However, there are examples where efficiencies can be gained in both areas. Indeed, a study in one Indian district showed that 90 % of the functioning pumps were less than 30 % efficient. Replacement of the pumps with correctly sized efficient units and conversion of a proportion of the flood irrigated fields to drip irrigation achieved overall efficiency improvements of 70 % in terms of energy and 60 % reduction in water usage (FAO 2012a). Thus, the identification of both the problems and solutions needs to be done on a case-by-case and regional basis.

The Millennial drought in Australia (1995 to 2010) is another example that galvanized action on more efficient water management of the Murray Darling Basin. The Chairman of the Murray Darling Basin Authority, Hon. Craig Knowles summarized the Murray Darling Basin Plan in a speech to the UN General Assembly in May 2013. He highlighted the positive outcomes that could be achieved with a bi-partisan approach by government, and multiple levels of government and community, involved in development and delivery of a scientifically-based approach to water management for agriculture, industry and the environment (Knowles 2013).

Energy Efficient Food Processing and Reducing Post-harvest Loss

Food processing operations are energy-intensive and provide multiple opportunities for reductions in energy intensity.

In the U.S. it is estimated that processing accounts for about one-third of the energy use in the U.S food system and each calorie of processed food consumes up to 1,000 calories of energy (Horrigan et al. 2002). Some estimates for energy inputs into processing various foods are 575 kcal/kg for canned fruit and vegetables, 1,815 kcal/kg for frozen fruits and vegetables, 15,675 kcal/kg for breakfast cereals and 18,591 kcal/kg for chocolate.

Horrigan and coworkers also reported that in 1997 the USDA estimated that the U.S. meat industry produced 1.4 billion tonnes of manure, equivalent to 5 tonnes of animal waste per capita. The manure output from the factory farms was too large to be absorbed by the local crop lands.

Solutions exist for these problems. Cogeneration of heat and power from manure and food waste can be used to power food processing plants, with excess energy exported to the grids. Increasing energy efficiency and integrating renewable energy could also assist with the sustainability of post-harvest processing of food. These approaches could be improved and/or implemented in food processing facilities in the developed and developing world and go some way to improving efficiency and sustainability. For example, in the U.S., the University of Minnesota's Minnesota Technical Assistance Program (MnTAP 2013) details energy efficiency opportunities for food processing facilities in Minnesota to realize 10–15 % energy savings by upgrading technology. They provide examples for a range of equipment in processing plants as follows:

Refrigeration systems consume a large amount of electricity in food processing facilities. However, savings up to 30 % of base electrical usage can be obtained through a variety of modifications, including improved insulation, tighter seals on doors and changes to the system controls and variable frequency drives for compressors or evaporator fans.

Pumps and fans can account for up to 15 % of the load in various facilities. When installed, pumps are often oversized to meet a maximum flow requirement; piping, valves, and the rest of the system may be undersized to contain costs. Both pumps and fans can be evaluated for energy saving opportunities during the design and installation phases, though there may be retrofitting opportunities available for large process pumps that run long hours as well as large cooling towers and HVAC systems. Oversized fans can also be slowed, resulting in a large decrease in energy consumption.

Combined Heat and Power (CHP) provides an opportunity to reduce the overall energy consumption in facilities by generating electricity on-site and recovering waste heat from the electrical generation for the production process. When a facility obtains its electricity from the local utility and generates thermal energy through the combustion of natural gas, the energy conversion process is only 33 % efficient. However, using CHP to produce electric energy on-site can result in 80 % efficiency.

CHP processes convert waste heat or steam into electrical power. The food industry produces biomass waste, which could be used as an alternative fuel source. CHP requires a large capital investment, which acts as a deterrent, but may be an attractive opportunity for food processing facilities that have high energy intensity, a flat year-round load profile, and high thermal to electric ratios.

As mentioned earlier, in developing countries most food-losses occur during harvest and storage. However, in the late-stage developing world, operational improvements with regards to logistics, supply chain management, effective storage and cold-storage as well as cost-effective processing has the potential to reduce post-harvest losses significantly (FAO 2012a). Low cost and energy-efficient cool storage systems have been developed and implemented in some parts of the world (Pelletier et al. 2011).

However, upgrading core infrastructure in the developing world will require significant capital expenditure to make significant inroads into the current level of food losses through spoilage. Concerted efforts by the National and Local governments in partnership with Industry and the local community are imperative.

References

Borre, L. 2013. Harmful algae blooms Plague Lake Erie again. Retrieved 4 Jan 2015, from http://voices.nationalgeographic.com/2013/04/24/harmful-algae-blooms-plague-lake-erie-again/.
FAO. 2012a. Energy-smart food at FAO, Food and Agriculture Organisation of the United Nations.
FAO. 2012b. Road to Rio: Improving energy use key challenge for world's food systems. Retrieved 4 Jan 2015, from http://www.fao.org/news/story/en/item/146971/icode/.
Greiff, J. 2015. No more sushi unless we role with farmed fish. *Australian Financial Review*. Sydney: Fairfax Media Publications Pty.
Horrigan, L., R.S. Lawrence, and P. Walker. 2002. How Sustainable agriculture can address the environment and human health harms of industrial agriculture. *Environmental Health Perspectives* 110(5): 445.
Knowles, C. 2013. Sustainable development and climate change: Practical solutions in the energy-water nexus. Retrieved 4 Jan 2015, from http://www.mdba.gov.au/media-pubs/mr/mdba-chair-speech-to-un.
Leaver, J.D. 2011. Global food supply: a challenge for sustainable agriculture. *Nutrition Bulletin* 36(4): 416–421.
MnTAP. 2013. Increasing energy efficiency in food processing facilities. Retrieved 4 Jan 2015, from http://www.mntap.umn.edu/food/energy.htm.

Pelletier, N., E. Audsley, S. Brodt, T. Garnett, P. Henriksson, A. Kendall, K.J. Kramer, D. Murphy, T. Nemecek, and M. Troell. 2011. Energy intensity of agriculture and food systems. *Annual Review of Environment and Resources* 36: 223–246.

Rayner, V., E. a. Laing and J. Hall. 2011. Developments in Global Food Prices. Reserve Bank of Australia Electronic Bulletin Publication.

Stone, R. 2011. On Lake Taihu, China moves to battle massive algae blooms. Retrieved 4 Jan 2015, from http://e360.yale.edu/feature/on_lake_taihu_china_moves_to_battle_massive_algae_blooms/2429/.

Chapter 6
Improving Sustainability in Agriculture

Feeding a growing world population will require a 60 % increase in food production by 2050, but we are not going to be able to meet that goal the way we did during the Green Revolution, relying on fossil fuels, a very different approach is required. Alexander Miller.

FAO Director-General for Natural Resources and the Environment, (FAO 2012).

Transforming our cities to a more sustainable and efficient consumption of resources require socio-technical approaches, starting with a concerted effort to foster community awareness and behavioural change for efficient consumption of water, energy and food. Exploiting the water-energy nexus in urban development, such as district-level tri-generation and the further utilisation of available heat for water disinfection and production of district-level reticulation of hot water, are simple cathartic initiatives to lead this transformation. Professor Tony Wong.

CEO, CRC for Water Sensitive Cities, (Wong 2013).

A new technology revolution is needed to meet the food production challenges of the next century. This will have to address the challenge of feeding a much larger global population, and of satisfying their nutritional needs in both quantity and quality of food. Future increases in food production will have less reliance than in the past on using increasing amounts of land, water and energy and on the exploitation of the environment. Much of the increased food production will be from developing countries. Collaboration between state and private sectors and between countries in research and development will be important in achieving this objective. J.D. Leaver, Royal Agricultural College (Leaver 2011).

As clearly articulated in the selected quotes above, it is widely recognized that the gains in crop production and productivity achieved over the last century are often accompanied by negative effects on the planet's natural resource base, highlighting the sustainability dilemma to meet food security and deal with the Food-Water-Energy nexus over the next century. Add to this the core issue of land availability which is succinctly summarized in a report from the Institution of Mechanical Engineers— Global Food, Waste Not Want Not (IME 2013). According to this report, land surface area of the planet is 14.8 giga hectares (Gha), of which about 10 Gha is capable of supporting productive biomass, i.e. excluding deserts, tundra, mountains etc. Global

M. Verma, *Energy Use in Global Food Production*,
SpringerBriefs in Food, Health, and Nutrition, DOI 10.1007/978-3-319-16781-7_6

food production currently uses 4.9 Gha of this to support one species—humans, and our pets. This leaves just over 50 % of the productive land area for the majority of the other land based species and ecosystems. Clearly, we need to find ways of expanding the amount of available food with minimal increase in cultivated land.

The world needs a new paradigm with two major focus areas.

1. The simplest and least cost way to produce more food is to **reduce the current level of loss and waste** from current production systems.
2. Innovation to deliver **sustainable intensification** of agriculture.

Minimizing Food Loss and Waste

Change is required at every level of the food supply chain from producers to consumers, and in every region developed and developing. The type of intervention or improvement and associated cost varies depending on the region.

In the developed world, food wastage by consumers post-purchase and that driven by wholesalers and retailers needs to be reduced. This is the most expensive loss to the planet currently, given the embodied energy and water in the food at this end of the supply chain.

Fortunately, there appears to be anecdotal evidence of change emerging slowly, driven by concerned citizens and a number of non-government agencies.

WRAP (Waste and Resources Action Programme) is a not-for-profit organization in the U.K. working on waste reduction and efficient resource utilization (WRAP 2013). They work with business, government and the community. WRAP have developed strategies applicable to all stages of the food chain in the U.K., and they have developed mechanisms to track food waste more accurately. Their latest survey shows that avoidable household food waste in the U.K. was cut by 21 % between 2007 and 2012, amounting to a saving to consumers of £13 billion over the five years. This demonstrates that consumer behavior can be changed and is a great start. There is a further opportunity to avoid an additional 4 million tonnes of avoidable food waste per annum. WRAP is now extending their activities to Europe through the Fusions program using social engineering.

Save Food Initiative (SaveFood 2014) is another example of international business and government agencies, partnering to develop solutions to minimize food waste. This initiative is a collaboration of the Messe Düsseldorf Group with FAO and UNEP. The Save Food Initiative aims at encouraging the dialogue on food losses between industry, research, politics and civil society. They count some of the world's largest agribusiness and food packaging companies as members. The Save Food Initiative launched the Think.Eat.Save campaign in 2014.

In Australia, Oz Harvest was set up 10 years ago as a perishable food rescue organization (OzHarvest 2014). They collect quality excess food from commercial outlets and deliver it, free of charge, to charities providing assistance to vulnerable people. Oz Harvest operates in several cities across eastern and southern Australia where their signature bright yellow vans can be seen redistributing food. Over the last 10 years Oz harvest states that it has delivered over 30 million meals and saved 10,000 tonnes of food from wastage. Oz Harvest is the Australian partner of the United Nations Think.Eat.Save campaign.

The U.S. has a nationwide Food Recovery Network that estimates it has donated over 600,000 pounds of food in the last 3 years (FoodRecoveryNetwork 2012). Started by students at the University of Maryland to reduce the food wasted at cafeteria and feed the hungry in the D.C. area, this organization has expanded to programs at 95 colleges around the U.S. in a little over 3 years.

In the developing world reduction of food loss at the producer and transportation level requires a range of approaches. On-farm storage needs to be improved at the small holder level, integrated pest management strategies need to be customized on a case-by-case basis and centralized cold storage and road transport logistics need to be improved. To significantly reduce food losses, major investment will be required in infrastructure at different stages of the food supply chain.

Sustainable Intensification of Agriculture and Food Production

The two-pronged approach required to achieve gains in this area are to (i) reduce energy-intensity of crops in the developed world without decreasing yield, and (ii) increase intensity of production in the developing world without significant increase in energy and water intensity.

Several researchers have published strategies for intensifying production sustainably in the developing world. The core message is that this will not be a one-size fits all. There are reports that traditional mixed farming approaches can be developed further and intensified in a sustainable manner in small holdings in the developing world (Herrero et al. 2010). Countries like India in the late-stage developing world have been concerned about sustainable intensification for some time and have reported on the energy efficiency of various crop rotation cycles for intensive forage production to provide grazing for livestock (Lal et al. 2003). Translation of these studies into practice at the farm gate and village level is being monitored.

In the developed world there is an increasing interest in 'conservation agriculture' and zero tillage approaches to reducing on-farm energy, and improved energy efficiency in the food processing industry. Co-generation of energy from waste biomass for use on the farm and by food processors is another approach that is being explored for reduction of energy intensity, without loss of productivity.

Case Studies

The section below expands on selected case studies from different regions to illustrate some of the strategies discussed above. Case study information is summarized from published books and reviews and it is beyond the scope of this work to critically analyse the source data. They are presented to highlight possible solutions for increasing sustainable agriculture and reducing the overall global energy input.

Reducing Post-Harvest Loss

FAO's Post-Harvest System

In the early-stage developing world post-harvest loss of cereal grains is significant. In Asia in 1997, post-harvest loss of the rice crop was estimated to be 14 %, or 77 million tonnes. Some stages in the post-harvest process, like drying and storage are more critical than others.

The FAO's rice post-harvest system concept is an efficient, modern approach that focuses on preventing post-harvest losses and ensuring the quality and safety of the rice crop during its processing and storage (Mejía 2004). The system includes procedures that add value to both primary and secondary rice products, as well as by-products.

The system includes small metal silos for storing grains at the household level varying in capacity from 100 to 4,000 kg. For a family of five people, a silo of 1 tonne capacity can maintain the quality and safety of rice for up to a year, thereby contributing significantly to household food security. A silo of this size costs about US$55 (2004 prices) and lasts for between 15 and 20 years. Small portable fan dryers assist with maintaining quality. The system also includes ways of processing and using rice by-products at the domestic level. For example, rice husks can be used first as fuel and then, after they have been burned, as fertilizer. Households can also produce rice pellets for feeding their own fish or selling to aquaculturists. These pellets are made out of flour, which is ground from broken grains and mixed with rice bran. Post-harvest technology plays an important role in the food security of households and communities.

Action Contre la Faim (ACF) also reported that the household metal silo is one of the key post-harvest technologies in boosting efforts for food security (Kiaya 2014).

Sustainable Intensification

There are many pilot initiatives centered around the introduction of sustainable agricultural practices, but technical documentation in this area is scarce. However, it seems that understanding of the key principles involved is growing around the world. These principles include (i) minimum soil disturbance, or no tillage at all; (ii) soil cover—permanent if possible; and (iii) crop rotation.

FAO defines conservation agriculture (CA) as follows: "CA is a concept for resource-saving agricultural crop production that strives to achieve acceptable profits together with high and sustained production levels while concurrently conserving the environment. CA is based on enhancing natural biological processes above and below the ground. Interventions such as mechanical soil tillage are reduced to an absolute minimum, and the use of external inputs such as agrochemicals and nutrients of mineral or organic origin are applied at an optimum level and in a way and quantity that does not interfere with, or disrupt, the biological processes."

Conservation Agriculture in Africa

Three case studies from Tanzania have been documented from the African Conservation Tillage (ACT) Network (Shetto and Owenya 2007). ACT is a pan-African association which encourages smallholder farmers to adopt conservation agriculture practices. The network involves private, public and non-government sectors; farmers; input suppliers and machinery manufacturers; researchers and extension workers—all dedicated to promoting conservation agriculture. The aim of specific short-term documented case studies was to provide results and outputs to CA practitioners, scientists and decision makers, to help improve planning and implementation of CA.

The Tanzania case study was carried out in three regions—Mbeya, Arumeru and Karatu. By way of background, agriculture in Tanzania is largely smallholder subsistence. Yields are generally low averaging <1 tonne/hectare, due to low and declining soil fertility, soil and water loss through erosion, and erratic rainfall. Increased livestock and human pressures have led to a collapse of conventional soil conservation systems and increased degradation through compacted soil, depletion of nutrients and organic matter and low water-holding and capacity and microbial activity. Conventional farming practices such as burning or removing crop residue and intensive tillage makes these problems worse. Promoting conservation agriculture in this region was seen to be valuable as a combination of crop and crop-livestock production practices had the potential to make the land more productive and improve the resilience of natural resources.

By the end of the two year project, the farmer-field training school groups increased from 31 to 44. The number of households in the project had increased from about 775 to over 1,200 and approximately 5,000 farmers had adopted some elements of CA through different organizations in the region.

Short-term benefits were (i) increase in crop yield, maize increased from 26 to 100 %, sunflower by 360 %; (ii) less labour needed as hand hoeing takes 3 people a day to plant one acre, while one person with a hand jab planter takes 3–4 h to plant an acre; and (iii) less labour for preparing the land from slashing, collecting and burning trash to just slashing.

Long-term benefits were (i) reduced soil erosion, lessening gullying and land degradation; (ii) improved soil fertility, structure and water-holding capacity; (iii) 2–6 times higher and stable yields and (iv) livelihoods and social interaction increased.

Key findings of obstacles to overcome, from these and other related studies in Africa (Shetto and Owenya 2007) were:

Tillage—many farmers do not go directly to no-tillage, relying on reduced tillage as an intermediate step, largely due to restricted access to no-till seeders.

Soil cover—producing sufficient biomass to cater for both adequate soil cover and provide grazing for livestock is a challenge. However, multiple-purpose cover crops like the field bean *Dolichos lablab* provide some success as the crop is able to provide food (grain and leaves are edible), income, forage and soil cover.

Weed control—remains a challenge when farming is done manually and in the early stages of CA adoption. As soils are not adequately covered, reducing tillage increases weeds. Both manual hoeing (four to six times), and herbicide use is not feasible due to cost, so more work is required to identify and promote suitable cover crops to achieve soil cover.

Equipment and inputs—reduced tillage implements such as rippers and no-till seeders were made available on an experimental basis to farmers. However, large scale and longer adoption of CA will require local manufacturing to keep costs down and the introduction of innovative access schemes like implement sharing or hire services along with rural financing schemes. As family labour becomes increasingly scarce, uptake of mechanization will need to increase.

Farmer Groups—A participatory approach with government institutions and publicly funded projects is essential, to enable farmer-field schools where groups of 10–30 farmers engage to experiment and learn CA principles and practices. Farmers need to be empowered and in control of the approaches they adopt.

Indigenous knowledge and innovative technology—some indigenous practice is compatible with the principles of CA, for example slash-and-mulch systems and cereal-legume intercropping. It is important to maintain a balance between imported innovative technology and local indigenous knowledge and the background reasons for certain practices in order to get better outcomes in the long run, as is illustrated in the next case study.

Agroecology 'Revolution' in Latin America

Agroecological initiatives essentially aim to transform industrial agriculture away from fossil-fuel reliance, using science and local knowledge to maximize production from smallholder mixed cropping farms. The core principles of agroecology include recycling nutrients and energy on the farm rather than relying on external inputs; enhancing soil organic matter and soil biological activity; diversifying plant species over time; integrating crops and livestock and optimizing productivity of the whole farming system rather than the yields of individual species.

Altieri and Toledo reviewed the spread of the agroecological paradigm across Latin America and its benefits for food production in the region (Altieri and Toledo 2011). Agroecology is knowledge-intensive and is based on techniques that are developed on the basis of farmers' knowledge and experimentation. Local communities experiment, evaluate and scale-up innovations through farmer-to-farmer extension approaches. This has resulted in a very active *Campesino a Campesino* (farmer-to-farmer) movement across a range of countries in South America, and has led to a revitalization of small farms to meet the regions food needs sustainably.

Given the biodiversity and range of ecosystems from Amazonian forests to the mountainous Andes in Latin America, agroecologists have long argued that modern farming methods will necessarily have to be rooted in the rationale of indigenous agriculture and build on this to design a biodiverse, sustainable, resilient and efficient agriculture. The peasant or small farm sector of about 65 million people with small farms (average size 1.8 ha) is very important for the supply of food in the region. This sector produces 51 % of the maize, 77 % of the beans and 61 % of the potatoes consumed in the region, for example in Brazil alone, 4.9 million family farms occupy 30 % of the total agricultural land in the country and produce 88 % of all the cassava and 67 % of all the beans. Agroecologists have shown that small family farms can be more productive than large farms if total output is measured rather than the yield of a single crop. For example in Brazil polyculture with maize and beans yielded a 28 % advantage, with minimal external inputs.

Brazil has had a dramatic expansion in agroecology since the 1980s. Agroecology is now seen as an emerging science and field of transdisciplinary knowledge influenced by social, agrarian and natural sciences.

Agroecology has also taken hold in Cuba over the last two decades, in part due to the sanctions and limited access to external inputs. One hundred thousand families or about 50 % of the independent farmers practice agroecological diversification and have increased their yields beyond industrial agriculture, such that they now produce over 65 % of the country's food with about 25 % of the land.

Other examples include the soil conservation practices introduced to Honduras with hillside farmers increasing their yields from 400 to 1200–1600 kg/ha. The introduction of velvet bean to fix nitrogen helped triple maize yields to 2500 kg/ha, while eliminating the need for herbicide use.

In Nicaragua, degraded land has been recovered, utilization of crop cover reduced mineral fertilizer use from 1,900 to 400 kg/ha with production costs 22 % lower than farmers using fertilizers and monoculture.

The highlands of Peru have >600,000 ha of highland terraces, mostly constructed in prehistoric times. The raised beds and canal systems are being revived to provide more sustainable and higher productivity yields.

The number of small farmers has increased in Latin America, and by exploiting environmental services derived from biodiverse ecosystems and using locally available resources with minimal external inputs, these farmers may be able to secure long term food security for the region in a sustainable manner.

Improving Energy Efficiency

Adoption of no-till Farming in Australia

It is now recognized that reduced soil disturbance through no-till conservation agriculture methods has led to greater profitability, sustainability and reduced environmental impact in the Australian cropping belt. Uptake of no-till cropping practices by grain growers in Australia is relatively recent. However, a survey by the Grains Research and Development Authority (GRDC) reported that in most regions 90 % of growers are using no-till systems to some extent (Llewellyn and D'Emden 2010).

The South Australian No-till Farmer's Association (SANTFA) in conjunction with the Conservation Agriculture Alliance of Australia and New Zealand (CAAANZ) commissioned the GRDC report. The study included 1,172 grain growers from 19 selected grain growing regions across the country. Fourteen of the selected regions show no-till farming practiced by 80–90 % of the growers. These high adoption areas appear to be plateauing at around 90 % adoption rates. The remaining 5 regions have adoption rates between 45 and 75 %. Some of the lower adoption areas appear to have plateaued, while others may reach 80 % adoption over time. Overall the proportion of adopters is expected to increase over the next five years, peaking at 90 % in most regions, with an average 70 % of crop sown with no-till by adopters. In Western Australia, almost 100 % of the crop sown is no-till. Growers around the country, who do not use no-till systems, tend to have significantly smaller cropping areas and a preference for managing livestock. They also have less use of paid cropping consultants. Interestingly, non-adopters also have a lower likelihood of having someone with a higher education involved with managing the farm.

Growers who used no-till did do because these practices (i) reduced fuel use and cost, and labour costs at seeding, (ii) improved soil conservation and (iii) improved soil moisture management. Once growers adopt no-till practices, very few cease using these systems. However, herbicide costs can have a negative impact on local tillage use. For example, when glyphosate prices increased, 21 % of no-till users reported increased use of some tillage.

This uptake of conservation agriculture in a developed country has occurred relatively rapidly over two decades, with adoption of no-till at just 10 to 20 % in 1994.

References

Altieri, MAa, and V.M. Toledo. 2011. The agroecological revolution in Latin America: Rescuing nature, ensuring food sovereignty and empowering peasants. *The Journal of Peasant Studies* 38(3): 25.

FAO. 2012. Road to Rio: Improving energy use key challenge for world's food systems. Retrieved 4 January 2015, from http://www.fao.org/news/story/en/item/146971/icode/.

FoodRecoveryNetwork. 2012. Food recovery network. Retrieved 5 January, 2015, from http://www.foodrecoverynetwork.org/about-us/our-work/.

Herrero, M., P.K. Thornton, A.M. Notenbaert, S. Wood, S. Msangi, H.A. Freeman, D. Bossio, J. Dixon, M. Peters, J. van de Steeg, J. Lynam, P. Parthasarathy Rao, S. Macmillan, B. Gerard, J. McDermott, Ca Seré, and M. Rosegrant. 2010. Smart investments in sustainable food production: Revisiting mixed crop-livestock systems. *Science* 327: 822–825.

IME. 2013. *Global food; waste not, want not.* London: Institution of Mechanical Engineering.

Kiaya, V. 2014. *Post harvest losses and strategies to reduce them.* New York: Action Contre la Faim (ACF International).

Lal, B., D.S. Rajput, M.B. Tamhankar, I. Agarwal, and M.S. Sharma. 2003. Energy use and output assessment of food-forage production systems. *Journal of Agronomy and Crop Science* 189(2): 57–62.

Leaver, J.D. 2011. Global food supply: A challenge for sustainable agriculture. *Nutrition Bulletin* 36(4): 416–421.

Llewellyn, Ra, and F.H. D'Emden. 2010. *Adoption of no-till cropping practices in Australian grain growing regions.* Australia: Canberra.

Mejía, D.J. 2004. *Rice post-harvest system: An efficient approach.* Rome: Food and Agriculture Organization of the United Nations (FAO).

OzHarvest. 2014. Counting on love. Retrieved 5 January, 2015, from http://www.ozharvest.org/.

SaveFood. 2014. Solutions for a world aware of its resources. Retrieved 5 January, 2015, from http://www.save-food.org/.

Shetto, R. and M. Owenya, eds. 2007. Conservation agriculture as practised in Tanzania: Three case studies. Nairobi, African Conservation Tillage Network: Centre de Coopération Internationale de Recherche Agronomique pour le Développement, Food and Agriculture Organization of the United Nations.

Wong, T. 2013. The urban water-energy-food nexus. CRC for Water Sensitive Cities. Retrieved 4 January, 2015 from http://watersensitivecities.org.au/.

WRAP. 2013. Household food and drink waste in the United Kingdom 2012. Retrieved, 4 January, 2015 from http://www.wrap.org.uk/content/household-food-and-drink-waste-uk-2012.

Chapter 7
The Way Forward

"What is at stake is the survival of our civilization and the habitability of the Earth. Or as one eminent scientist has put it, *'the pending question is whether an opposable thumb and a neocortex are a viable combination on this planet'*." (Gore 2006).

In the article quoted above, former U.S. Vice President Al Gore issued a call for urgent action on the threats and dangers of climate change. While anthropogenic climate change will undoubtedly impact agriculture and food security negatively, it will only exacerbate the growing pressures from an expanding and increasingly affluent global population. Clearly, limiting global population size is the key to sustainable and equitable access to limited interconnected resources.

Thomas Malthus articulated the Malthusian population trap as the threshold population level at which population increase was bound to stop because life-sustaining resources, which increase at an arithmetic rate, would be insufficient to support human population which increases at a geometric rate (Todaro and Smith 2011). The Malthusian trap clearly relates to isolated populations like Easter Island and the Mayan Civilization that collapsed due to resource limitations (Diamond 2005). Malthus pre-dated the Industrial Revolution and he did not foresee the "Green Revolution", where access to cheap fossil-derived energy has allowed resource limitations to be bypassed in many instances. For example, the pumping of fresh water from deep aquifers is highly energy intensive and allows withdrawal of water much faster than the reservoirs can be replenished naturally. Only when energy is priced to include negative externalities, will resource limitation costs start to be felt more acutely.

Therefore, Malthus's basic principle that human populations must self-regulate for a sustainable future with a reasonable standard of living per capita, still holds true and policies relating to food and water security need to be accompanied by policies relating to availability of affordable contraception and family planning (Pimental and Pimental 2000, Hugo 2012). As the distinguished economist Jeffery Sachs states completing the demographic transition is urgent and requires renewed effort on the education of girls and empowerment of women, particularly in the

© The Author(s) 2015
M. Verma, *Energy Use in Global Food Production*,
SpringerBriefs in Food, Health, and Nutrition, DOI 10.1007/978-3-319-16781-7_7

early-stage developing world (Sachs 2008). Sachs sets out the factors important in leading to rapid or slow decline in fertility rates and the policy initiatives required to stabilize population globally.

Turning to climate change, the recent IPCC report (IPCC 2014) makes the point that cooperative responses, including international cooperation are required to effectively mitigate GHG emissions and adapt to the impact of climate change. Outcomes seen as equitable can lead to more effective cooperation. The report states that "Transformations in economic, social, technological, and political decisions and actions can enhance adaptation and promote sustainable development (*high confidence*). At the national level, transformation is considered most effective when it reflects a country's own visions and approaches to achieving sustainable development in accordance with its national circumstances and priorities."

There are expected to be aggregate economic costs of mitigation, estimates of which vary depending on the assumptions employed. However, in a scenario where all countries begin mitigating immediately, using a single global carbon price and all alternative technologies are employed, warming is likely to be limited to below 2 °C, and the impact on consumption is estimated to be an annualized reduction of growth by 0.04–0.14 % points over the century, relative to annualized consumption growth in the baseline between 1.6 and 3 % per year (IPCC 2014). In order to achieve effective mitigation, decarbonization of electricity generation and reduction in energy intensity and demand are key measures, consistent with the Water-Food-Energy nexus approach discussed in this book. IPCC sets out a range of policies including effective cap and trade, as well as carbon and fuel taxes, complemented by regulatory approaches for countries to achieve GHG mitigation targets. Effective climate change mitigation is expected to have an overall positive outcome for food security (IPCC 2014).

The World appears to be making some halting steps towards developing a global mitigation approach. The recent U.S. and China announcement of 'intentions' toward GHG mitigation (Adler 2014) is a step in the right direction, since these two countries are jointly responsible for about a third of global emissions. The Global Climate Change Conference in Lima (COP20) in December 2014 made some important steps towards effective mitigation, but critical decisions around targets were postponed to the COP21 meeting in Paris in December 2015 (Upton 2014). All informed and concerned citizens should be encouraging their respective governments to make binding commitments towards shouldering their share of mitigation efforts in the forums leading to the Paris talks.

In parallel with efforts to mitigate and adapt to climate change, the global approach to agriculture needs a paradigm shift with adoption of key technical, policy and planning tools that have been developed over recent years. The Water-Energy-Food Nexus planning and decision-support framework (Bizikova et al. 2013), developed by the International Institute for Sustainable Development provides a good place-based approach that should be adopted by key food producing regions.

In addition, conservation agriculture and related agroecology approaches are being adopted by farmers in developed and developing countries, with encouraging implications for lowering energy intensity of food production, while addressing global food security issues. Pretty and co-workers summarized the improvements in food production from 208 projects in 52 developing counties and concluded with cautious optimism, that the evidence indicates that productivity can increase (up to 93 %) steadily over time if natural, social and human capital assets are accumulated (Pretty et al. 2003). The authors conclude that national policy reforms and better markets could expand these benefits significantly. However, a transition towards sustainable agriculture in the early-stage developing world will require significant external help and investment. It has been suggested that agricultural investment portfolios funded by the G8 countries, should include payments for protecting water, carbon, biodiversity, and other global goods and ecosystem services where appropriate.

In the developed world, conservation agriculture is starting to make a contribution where adopted, but there remains a need for agricultural research and development to be re-invigorated in many countries with a focus on sustainable intensification of food production (Leaver 2011).

The lowest cost and lowest impact way to increase net available food and reduce energy use in food production is undoubtedly through reduction of food wastage in the developed world. Local NGOs like OzHarvest, Food Recovery Network and WRAP are starting to make a difference by raising consumer awareness and responsibility. In some cases this has required legislative change. For example, in Australia in 2005, Ronnie Kahn (founder of OzHarvest) successfully lobbied State governments to allow food donations to charitable organizations by changing legislation to protect food donors from liability for their donation (OzHarvest 2014). Likewise, the FAO's Think.Eat.Save initiative is taking a holistic approach to the issue of food waste. The prime area to address in developed countries, however, is the fact that under current market conditions, many staple foodstuffs are regarded as low-cost commodities and, as such, rarely receive the focus on waste that they deserve. Therefore, to see significant improvements in this area, developed countries may need to set targets for waste reduction. This could be done informally by a partnership of key stakeholders, as has been done in Switzerland (Think.Eat.Save 2014), or more formally by regulation and government policy (GO-Science 2011). Demand for more resource intensive food needs to be curtailed, through campaigns to inform consumer choices in the developed world and the late-stage developing world.

Cities can also lead the way towards economically, socially and environmentally sustainable societies through a holistic approach to urban planning and management. Megacities, if planned and managed properly could reduce food waste and reduce energy intensity and transportation costs for food on a per capita basis (United Nations 2014).

In early-stage developing countries, reducing food losses and wastage requires development of an integrated approach including knowledge transfer, educational programs, reduction in free trade obstacles (that result in damage to fragile produce at border crossings) and investment in infrastructure (IME 2013). Financing institutions like the U.N., World Bank and IFC need to coordinate efforts to develop innovative funding schemes for grain storage and transportation. The scale and cost of these projects is large and they will probably require effective public-private partnerships (IME 2013).

A number of issues in the Water-Energy-Food nexus arise because the economic value of these goods does not include 'externalities', and finding a method or tool to address this issue globally, is proving to be a major hurdle for moving towards sustainability.

A colorful illustration of the principle comes from the concept of a $200 Big Mac (Patel 2009). The author asserts that this would be the price for the product, if all the costs to society for producing a Big Mac were counted, including GHG emissions, feed subsidies, social subsidies and public health costs.

In fact, economists seem to have had more success than scientists at communicating the case for GHG mitigation. Lord Stern in the UK (Stern 2006) and Ross Garnaut in Australia (Garnaut 2008) succeeded in making the general public and politicians aware of climate change and related potential negative impacts, information that scientific experts had been aware of for several decades previously. The economists argued that there were market failures at work that could only be rectified if negative externalities were priced into the economy. Their recommendations have not been picked up quite as comprehensively as one may have hoped, judging by the current lack of a global price on carbon.

Help may be at hand from an unexpected source—accountants! A recent book (Gleeson-White 2014) describes a revolution taking place in accounting—Integrated Reporting. In late 2013, the International Integrated Reporting Council (IIRC) published a framework of a new way of business thinking and reporting; the International Integrated Reporting Framework (IR). This framework includes reporting against six 'capitals' used to produce goods and services: financial, manufactured, intellectual, human, social and relationship, and natural capital. As this approach is more generally adopted, it will provide a mechanism for including information about externalities, both negative and positive, in the goods and services we consume. At this stage, financial capital retains primacy over the other capitals but the business community is being encouraged to consider the impact of all the six capitals in their value creation, even those they do not own or those that are beyond their control, such as carbon emissions. The IIRC calls this the prism through which organizations can assess and report the extent to which they are creating, diminishing or destroying value over time.

Natural capital is defined as the renewable and non-renewable environmental resources and processes that provide goods and services that support the organization's past, present and future prosperity, including air, water, minerals, forests, biodiversity and ecosystem health. Already, some major global companies have developed methodologies to report their environmental impact across their entire supply chain. As this initiative expands, and is adopted more generally by the food industry, an appreciation for the related energy use, its cost and broader impact on the planet may be better appreciated.

It is fitting to conclude with the words of David Pimental, Professor of ecology and agricultural sciences at Cornell University, who was instrumental in raising awareness about embedded energy in food production in the early days of this work.

> Increases in food production, per hectare of land, have not kept pace with increases in population, and the planet has virtually no more arable land or fresh water to spare. As a result, per-capita cropland has fallen by more than half since 1960, and per-capita production of grains, the basic food, has been falling worldwide for 20 years (Pimental and Wilson 2004).

To improve the growing imbalance between population numbers and food supply, humans should actively conserve croplands, fresh water, energy and biological resources. Populations in developed countries could contribute by reducing their high consumption of resources. The development of appropriate, safe technologies holds the promise of improving food production.

As the human population increases, the right to freedom from malnutrition, hunger, poverty, and diseases is gradually eroded. Freedom to enjoy open space and treasured natural environments also is infringed. If we are not brave enough to limit our numbers, nature will impose its own limits on us (Pimental and Pimental 2000).

References

Adler, B. 2014. New U.S.–China climate deal is a game changer. Retrieved 22 Jan 2015, from http://grist.org/climate-energy/new-u-s-china-climate-deal-is-a-game-changer/.

Bizikova, L., D. Roy, D. Swanson, H. D. Venema and M. McCandless. 2013. The water–energy–food security nexus: Towards a practical planning and decision-support framework for landscape investment and risk management, International Institute for Sustainable Development.

Diamond, J. 2005. *Collapse. How societies choose to fail or survive*. Camberwell, Australia: Penguin Group.

Garnaut, R. 2008. The Garnaut Climate Change Review. Cambridge University Press, Melbourne, Australia.

Gleeson-White, J. (2014). *Six capitals.The revolution capitalism has to have—or can accountants save the planet?* Allen and Unwin: Sydney, Australia.

Gore, A. 2006. The moment of truth; time to get real: global warming is the problem—the biggest problem. It's not a matter of "When?" any longer. It's here. Vanity fair.

GO-Science. (2011). Foresight. The future of food and farming. Final project report. The Government Office for Science, London.

Hugo, G. 2012. "Challenge 3: Balancing population growth and resources. Retrieved 22 Jan 2015, fromhttp://theconversation.com/challenge-3-balancing-population-growth-and-resources-7489.

IME. 2013. Global food; waste not, want not. London: Institution of Mechanical Engineering.

IPCC (2014). Climate change 2014 synthesis report. Report of the Intergovermental Panel on Climate Change. Retrieved 3 Jan 2015, from http://www.ipcc.ch/report/ar5/.

Leaver, J.D. 2011. Global food supply: A challenge for sustainable agriculture. *Nutrition Bulletin* 36(4): 416–421.

OzHarvest. 2014. Counting on Love. Retrieved 5 Jan 2015, from http://www.ozharvest.org/.

Patel, R. 2009. *The value of nothing: How to reshape market society and redefine democracy*. Picador Books.

Pimental, D. and A. Wilson. 2004. World population, agriculture and malnutrition. *World Watch Magazine*, from http://www.worldwatch.org/node/554.

Pimental, D., and M. Pimental. 2000. Feeding the world's population. *BioScience* 50: 387.

Pretty, J.N., J.I.L. Morison, and R.E. Hine. 2003. Reducing food poverty by increasing agricultural sustainability in developing countries. *Agriculture, Ecosystems and Environment* 95: 217–234.

Sachs, J. 2008. *Commonwealth. Economics for a crowded planet*. London, England: Penguin Group.

Stern, N. 2006. Stern review on the economics of climate change. HM Treasury, London.

Think.Eat.Save. 2014. "Think.Eat.Save reduce your footprint. Retrieved 5 Jan 2015, from http://www.thinkeatsave.org/.

Todaro, M.P., and S.C. Smith. 2011. *Economic development*, 11th ed. Harlow, England: Addison-Wesley.

United Nations, D. 2014. *World urbanisation prospects, 2014 revision*. New York: United Nations.

Upton, J. 2014. Climate accord struck in lima; key decisions postponed. Retrieved 22 Jan 2015, from http://www.climatecentral.org/news/climate-accord-struck-in-lima-18443.